Hind A. Al-Abadleh
Atmospheric Aerosol Chemistry

Also of interest

Atmospheric Chemistry.
A Critical Voyage Through the History
Möller, 2022
ISBN 978-3-11-073739-4, e-ISBN 978-3-11-073246-7

Electrochemical Carbon Dioxide Reduction
Perry, 2021
ISBN 978-1-5015-2213-0, e-ISBN 978-1-5015-2223-9

Methods for Scientific Research.
A Guide for Engineers
Prasad, 2022
ISBN 978-3-11-062529-5, e-ISBN 978-3-11-062530-1

Instrumental Analysis.
Chemical IT
Schlemmer G, Schlemmer J, 2022
ISBN 978-3-11-068964-8, e-ISBN 978-3-11-068966-2

Atmospheric Pressure Plasma.
Methods and Industrial Applications
Ananth, 2022
ISBN 978-3-11-064026-7, e-ISBN 978-3-11-064036-6

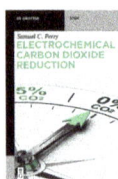

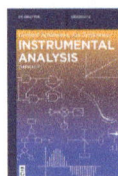

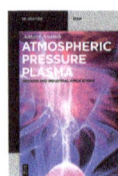

Hind A. Al-Abadleh

Atmospheric Aerosol Chemistry

State of the Science

DE GRUYTER

Author
Hind A. Al-Abadleh, Ph.D.
University Research Professor
Department of Chemistry and Biochemistry
Wilfrid Laurier University
Waterloo, ON N2L3C5
CANADA
519-884-0710 ext. 2873 (phone)
halabadleh@wlu.ca (e-mail)

ISBN 978-1-5015-1936-9
e-ISBN (PDF) 978-1-5015-1937-6
e-ISBN (EPUB) 978-1-5015-1256-8

Library of Congress Control Number: 2021953384

Bibliographic information published by the Deutsche Nationalbibliothek
The Deutsche Nationalbibliothek lists this publication in the Deutsche Nationalbibliografie;
detailed bibliographic data are available on the Internet at http://dnb.dnb.de.

© 2022 Walter de Gruyter GmbH, Boston/Berlin
Cover image: © Pattadis Walarput/iStock/Getty Images Plus
Typesetting: Integra Software Services Pvt. Ltd.
Printing and binding: CPI books GmbH, Leck

www.degruyter.com

Preface

The lower atmosphere, known as the troposphere, contains a large concentration of minute airborne particles, which together with the gas surrounding them are referred to as atmospheric aerosols. Some aerosols get emitted directly from sources like biomass burning, ocean wave action, dessert storms, and human activities, and others are formed in the atmosphere through gas phase chemistry that leads to particle nucleation and growth. The combination of air pollution and climate change is worsening air quality through higher concentrations of ground-level ozone and inhalable fine and ultrafine particles. Hence, research in atmospheric chemistry in the era of climate change demands innovative and integrated approaches to thinking at multiple scales, all phases of matter, and at the interface of different phases of matter.

The latest Intergovernmental Panel on Climate Change (IPCC) report released in summer 2021 associated higher certainties with humans' role in causing climate change, with no region in the world being spared from experiencing its impacts in terms of more intense and frequent weather extremes, wildfires, flooding, and drought. Atmospheric and climate models are developed based on data about atmospheric composition and dynamics. These models are validated by reproducing historical data from field measurements, and also incorporate bulk and interfacial reaction rates and mechanisms per available literature. However, aerosols' representation in these models is still far from complete. This is because aerosol physicochemical properties change during their atmospheric residence time, which can last up to two weeks.

The STEM book *Atmospheric Aerosol Chemistry: State of the Science* outlines major research findings to date in aerosol chemistry and advances in analytical tools used in laboratory settings for studying their surface and bulk reactivity. These processes take place at the surface of the particles/droplets or within the aerosol condensed phase. This book is organized into five main chapters that start with a general introduction in Chapter 1. "Instrumentation for measuring aerosols physical and chemical properties" is the focus of Chapter 2 followed by "Physical properties of aerosols" in Chapter 3. Chapters 4 and 5 highlight specific examples on "Interfacial aerosol chemistry" and "Bulk aqueous phase chemistry relevant to cloud droplets", respectively. Each chapter cites references closely related to the topics presented which gives the reader the opportunity to further explore a given concept or reaction at a much deeper level.

The writing of this book came at a time when exciting research questions enabled by advances in instrumentations were taking place among atmospheric chemistry research groups around the world. I hope this book would serve as the starting point for early career researchers eager to learn about atmospheric aerosols and pursue further research opportunities to advance this branch of knowledge. I am grateful to all the students, colleagues, and senior researchers whom I met in person, worked with,

https://doi.org/10.1515/9781501519376-202

or read their papers for their dedication and enthusiasm in applying fundamental chemistry concepts to atmospheric aerosol research. The science you produced inspired me to choose this field of study, and I am confident it will inspire many generations to come.

Hind A. Al-Abadleh, Ph.D

Contents

Chapter 4
Interfacial aerosol chemistry —— 121

Chapter 1
General introduction

Atmospheric composition of the lower atmosphere, known as the troposphere, comprises gases and minute airborne particles commonly referred to as aerosols. These particles have complex chemical composition and physical properties, and they impact our lives and the climate in a number of ways. Reduced visibility seen on hazy days and respiratory health problems are just two examples of aerosols impact on regional/ personal scales. On a global scale, aerosols contribute to climate change as they influence the amount of sunlight reaching the Earth's surface, alter properties of clouds, and provide media for chemical reactions in the atmosphere. Computer simulations of the climate give us the capability of quantifying the magnitude of aerosol contribution to climate change. Yet representation of aerosols in these simulations is still inadequate mainly because aerosols are complicated in nature and their reactivity and properties change with time while suspended in air. Thus, it is imperative to address scientific questions related to aerosols through experimental studies to improve our understanding of the impact of aerosols on our climate system. In this chapter, a general introduction is provided on concepts used in atmospheric aerosol science in relation to aerosol sources, properties, climate impact, and chemical processing.

1.1 Definition, sources, and properties

The IUPAC definition of atmospheric aerosols is "Mixtures of small particles (solid, liquid, or a mixed variety) and the carrier gas (usually air); owing to their size, these particles (usually less than 100 pm and greater than 0.01 pm in diameter) have a comparatively small settling velocity and hence exhibit some degree of stability in the Earth's gravitational field. An aerosol may be characterized by its chemical composition, its radioactivity, the particle size distribution, the electrical charge and the optical properties" [1]. The *IUPAC Gold Book* defines aerosols as "dispersions in gases. In aerosols the particles often exceed the usual size limits for colloids. If the dispersed particles are solid, one speaks of 'aerosols of solid particles', if they are liquid of 'aerosols of liquid particles'. The use of the terms 'solid aerosol' and 'liquid aerosol' is discouraged. An aerosol is neither 'solid nor liquid', but if anything, gaseous" [1].

In the lower troposphere, atmospheric aerosol particles originate from primary sources and secondary processes [2–8]. Aerosol particles from primary sources include mineral dust, sea spray, terrestrial primary biological aerosol particles (bioparticles, for short, such as fungal spores and pollen), and primary organic aerosol (such as brown carbon and black carbon). An emerging class of "unconventional"

https://doi.org/10.1515/9781501519376-001

mineral dust is the one produced from rapid urbanization and industrialization, particularly in the developing countries. This class is referred to as anthropogenic fugitive, combustion, and industrial dust (AFCID) [9]. AFCID largely contributes to emissions of fine particulate matter (PM2.5), which are known to be harmful to human health [10–12]. Global fluxes and mass loadings of AFCID have been severely underrepresented in regional and global models despite having surface PM2.5 measurement networks for emission inventories (see Surface Particulate Matter Network, https://www.spartan-network.org). Philip et al. [9] included AFCID emissions in a global simulation using GEOS-Chem. Figure 1.1 shows the annual mean concentration of PM2.5 for 2014–2015 total dust (top panel), natural mineral dust (middle panel), and AFCID (bottom panel) as simulated with the GEOS-Chem model. Also shown in the top panel are circles for different locations that compare the mean measured PM2.5 from the Spartan Network in 2013–2015 (inner circles) with simulated values (outer circles). It was estimated that 2–16 µg m^{-3} of AFCID increases PM2.5 concentrations across East and South Asia [9]. As noted by the authors, this concentration of simulated AFCID is comparable to that of natural mineral dust over parts of Europe and Eastern North America [9].

Secondary processes refer to particle formation from reactions in the atmosphere among gas-phase inorganic and organic precursors leading to the formation of ammonium nonsea salt sulfates, nitrates, and secondary organic aerosol (SOA) [13–16] from ammonia, sulfur-containing gases such as sulfur oxides, nitrogen oxides, and volatile organic compounds (VOCs) of biogenic [17] and anthropogenic [18] origins, respectively. Also, SOA may be formed from secondary condensed phase reactions during long-range transport leading to particle growth. The organic component in the fine aerosol particle fraction (diameter <1 µm) contributes to more than 50% of the aerosol mass [16]. Residence time of aerosol particles ranges from hours up to 10–15 days during which they undergo long-range transport over thousands of miles [19]. They are removed from the atmosphere mainly via sedimentation, dry and wet deposition. As a result, atmospheric aerosol particles contribute to the biogeochemical cycles of nutrients such as nitrogen, iron, phosphorous, and other transition metals [20, 21]. Table 1.1 lists additional details about each type of these aerosols.

Atmospheric aerosol particles have complex physical and chemical properties that evolve over time and govern their lifetime in the atmosphere and impact on the climate system. These properties also govern their biological and toxicological impacts, which are of importance to understand and quantify aerosol effects on ocean productivity and human health, respectively. Table 1.2 lists the definition of major keywords used in aerosol characterization and chemistry. Chapters 2–5 provide more details on these properties with specific examples.

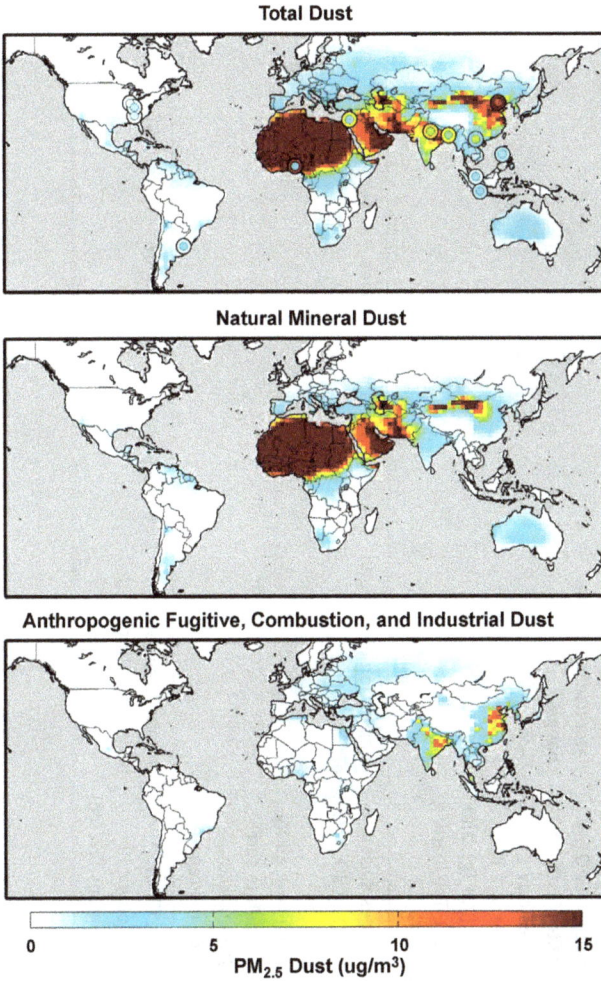

Figure 1.1: Annual mean (2014–2015) concentration of PM2.5 total dust (top panel), natural mineral dust (middle panel), and anthropogenic fugitive, combustion, and industrial dust (bottom panel) simulated with the GEOS-Chem model. Colored concentric circles in the top panel denote SPARTAN-measured campaign-mean (2013–2015) PM2.5 dust concentration (inner circle) and the coincident simulated value (outer circle). Reproduced from reference [9]. © The Author(s) 2017. Published by the Institute of Physics (IOP) Science Publishing.

Table 1.1: Key characteristics of atmospheric aerosol particles per type.

Type	Global emission (Tg year^{-1})	Atmos. loading (Tg)	Chemical composition	Source/formation mechanism	Size distribution	Lifetime[a]	Ref.
Mineral dust	1,840 1,000–4,000	19.2	Clay, oxides of alkali, alkaline earth, and transition metals, and carbonate. Latest estimates in ref. [22] of the top oxides by atmospheric loadings are quartz (SiO_2, 4.1 Tg) and hematite (Fe_2O_3, 0.2 Tg). The top three clay minerals are illite (4.2 Tg), montmorillonite (2.8 Tg), and kaolinite (2.2 Tg). The loading of calcite is 1.3 Tg.	Wind erosion, soil resuspension, some agricultural practices and industrial activities related to land use, water use, or climate change	Coarse and super-coarse mode with a small accumulation mode	1 d to 1 w[b]	[23] [2]
AFCID	13.1		Coal fly ash (oxides such as SiO_2, Al_2O_3, Fe_2O_3, CaO, K_2O, and MgO, as well as aluminosilicates) [24], oil fly ash [25] (ferric sulfate salt [$Fe_2(SO_4)_3 \cdot 9(H_2O)$] and nanosized Fe_3O_4 aggregates).	Combustion of coal and oil	Accumulation and fine modes (<2.5 µm)	w–d	[9, 26]
Sea spray	16,600 1,400–6,800^{c}	7.52	Sea salt (Na, Mg, Cl, Br, I) Marine POA in biologically active oceanic regions [27, 28]	Oceans (wave action, bubble bursting) [29]	Coarse and accumulation mode	1 d to 1 w[b]	[23] [2]

Species	Global burden	Ratio	Composition	Sources	Size modes	Lifetime[a]	Refs
Sulfate	179	1.99	S, O	Primary: marine and volcanic emissions. Secondary: oxidation of SO_2 and other S gases	Primary: Aitken, accumulation and coarse modes. Secondary: nucleation, Aitken, and accumulation modes	~1 w	[2, 23]
Nitrate	100		N, O	Oxidation of NO_x (x = 1, 2)	Accumulation and coarse modes.	~1 w	[2, 30]
POA (includes BrC) Anthrop. Biomass burning	96.6 6.3–15.3 29–85.3	1.7	Aromatic with diverse range of functional groups and molecular weights.	Combustion, forest fires, and direct sources from the biosphere.	Aitken and accumulation modes. Aged OA: accumulation mode. Freshly emitted BrC: 100–400 nm	1 w	[23] [2, 18] [2]
Black carbon	11.9 3.6–6.0	0.24	Aromatic with graphite-like structure	Combustion	Fresh: <100 nm Aged: accumulation mode	1 w to 10 d	[23] [2, 18]
SOA From biogenic VOCs	20–380		Small-to-medium-molecular-weight aromatic, aliphatic oligomers with diverse range of functional groups including nitrates, sulfates, iron.	Reactions among biogenic and combustion precursors, inorganic constituents, and oxidants in the gas and condensed phases,	Nucleation, Aitken, and accumulation modes,	1 w	[2, 31–34]
Terrestrial PBAPs	50–1,000		Lipids, proteins, sugars, enzymes, viruses, nutrients (N, P)	Terrestrial ecosystems [27]	Mostly coarse mode	1 d to 1 w[b]	[2, 20]

Notes: [a] Tropospheric. d, day; w, week. [b] Depending on size. [c] The range is 2–20 Tg year^{-1} including marine POA. PBAP, primary biological aerosol particles; POA, primary organic aerosol; BrC, brown carbon.

Table 1.2: Definition of selected terms used in atmospheric aerosol chemistry.

Term	Definition	Ref.
	Physical properties	
Albedo	IUPAC definition is *The fraction of the energy of electromagnetic radiation reflected from a body (or surface) relative to the energy incident upon it. The reflection of light from a surface is, of course, dependent on the wavelength of the light, the nature of the surface, and its angle of incidence with the surface. The term albedo usually connotes a broad wavelength band (visible, ultraviolet, or infrared), whereas the terms reflectivity and spectral albedo are used to describe the reflection of monochromatic (single wavelength or small band of wavelengths) radiation.*	[1]
Condensation nuclei (CN)	IUPAC definition is *A particle, either liquid or solid, or an ion upon which condensation of water vapour (or other substances) begins in the atmosphere. Condensation nuclei are usually very small hygroscopic aerosols (0.001 to 0.1 pm in diameter), but these are not as abundant as the smaller particles. The number of CN which are active (initiate condensation) in a given air mass may be a function of the relative humidity. Usually CN are counted as the active nuclei at about 300% relative humidity, while cloud condensation nuclei (CCN) are counted as the number of active nuclei at relative humidity less than or equal to 102%.*	[1, 35]
Hygroscopicity	Affinity of particles to gas-phase water is controlled by its chemical composition. This property influences the scattering efficiency of aerosols and the kinetics of their surface and multiphase reactions. It is quantified through changes in mass, diameter of particles, or surface coverage of water due to water vapor uptake.	[36]
Ice heterogeneous nucleation	An activated process by which surfaces or sites on aerosol particles (i.e., ice nuclei, IN) with a wide range of chemical composition, temperature, and relative humidity provide favorable conditions for nucleating ice from gas or liquid water.	[37, 38]
Mass concentration	Mass of particulate matter per unit volume of air ($\mu g\ m^{-3}$)	[1]

Table 1.2 (continued)

Term	Definition	Ref.
Mixing state	Distribution of properties across a population of particles within the aerosol. In the case of chemical mixing states, the term refers to the distribution of primary and secondary chemical species across individual particles within the particle population of aerosol with no physical properties. Physicochemical mixing state refers to the chemical speciation and physical properties of each particle. Often mixing state is described in terms of external and internal mixtures to refer to the number of chemical species per particle in an aerosol population: external mixed aerosol means each particle contains only one pure species, while internal mixed aerosol means particles that contain equal amounts of all chemical species. Freshly emitted particles are more externally mixed, where atmospheric aging leads to internal mixing. A mixing state index was developed to quantify the ratio of the average per particle species diversity and the bulk population species diversity D, both of which are based on information-theoretic entropy measures.	[39–42]
Morphology	A general term that reflects apparent shape, size, and topology of aerosol particles from optical and electron microscopy images. Examples of morphology include homogeneous, phase-separated core–shell and partially engulfed.	[43–45]
Number concentration	Total aerosol concentration (N_0) is the number of particles per m^3 obtained by integrating the number of particles per unit volume whose radius lies between r and $r + dr$: $N_0 = \int n(r)dr$	[7]
Optical properties	Ability to scatter and absorb radiation in the solar actinic region ($\lambda > 290$ nm). Near UV and visible-light absorbing organic carbon are referred to as "brown carbon."	[46, 47]
Phase	IUPAC definition is *In chemistry, a physically distinct, homogeneous portion of a heterogeneous mixture.*	[1]
	A physical state of matter, which for atmospheric aerosols, can be solid, semi-solid, or liquid. Mixed phases of aerosols exist in multicomponent aerosol systems containing organic, inorganic, and water components.	[48]

Table 1.2 (continued)

Term	Definition	Ref.
Radiative forcing due to aerosol–radiation interactions (RFari)	Quantification of the net radiative effect (in W m^{-2}) resulting from the scattering and absorption of shortwave and longwave radiation by atmospheric aerosols.	[7]
Salting out	Decrease in the solubility of a nonelectrolyte with increasing concentration of added electrolyte (i.e., salts), which is quantified by k_s, the salting constant in the Setchenov equation.	[49, 50]
Size distribution	IUPAC definition is *The size of the liquid or solid particles in the atmosphere usually extends from >0.01 to <l00 µm in diameter. In the earth's atmosphere the distribution function which describes the number of particles as a function of diameter, mass, or surface area of the aerosol can be determined reasonably well with modern instrumentation.*	[1]
	Particle size: To describe the size of liquid or solid particles (aerosol) the average or equivalent diameter is used. For nonspherical particles collected in an impactor, for example, the aerodynamic diameter of a particle of arbitrary shape and density refers to the size of a spherical particle of unit density that would deposit on a given impactor surface.	
	It is divided into four modes: nucleation and Aitken modes (diameter, d, from a few nm to 0.1 µm), accumulation mode ($0.1 < d < 1$ µm), and coarse mode ($d > 1$ µm).	[7]
Supersaturation	IUPAC definition is *In meteorology, supersaturation of an air mass with respect to H_2O vapour is of special interest. It is the saturation ratio minus one, or the percent supersaturation is the percent relative humidity minus 100.*	[1]
Viscosity	A measure of the resistance of a fluid to flow. The viscosity coefficient (η) is most commonly defined as dynamic (or absolute) viscosity, which has units of Poise (P) or Pa·s, or kinematic viscosity, which is dynamic viscosity divided by the density and has units of Stokes (St) or m^2 s^{-1}.	[51, 52]

Table 1.2 (continued)

Term	Definition	Ref.
Volatility	Partitioning of the aerosol organic component (i) between the particle and gas phases quantified through the effective saturation concentration (C_i*) in mass concentration units ($\mu g\ m^{-3}$) calculated from the saturation vapor pressure and the appropriate activity coefficient for the organic mixture. The aerosol volatility distribution under typical atmospheric conditions is "nonvolatile" (C_i* $< 0.1\ \mu g\ m^{-3}$), "semivolatile" (SVOC, $0.1\ \mu g\ m^{-3} < C_i$* $< 1,000\ \mu g\ m^{-3}$), and "intermediate volatile" (IVOC, $1,000\ \mu g\ m^{-3} < C_i$* $< 100,000\ \mu g\ m^{-3}$).	[53, 54]
	Chemical properties	
Aerosol acidity (pH)	Aerosol acidity is the quantification or estimation of hydronium ion (H^+) activity in liquid aerosol particles using either calibrated analytical methods suitable for the sample amount or theoretical model calculations. This property depends on the amount of liquid water and aerosol chemical composition in equilibrium with the gas phase.	[55–57]
Composition	Chemical identification and quantification of elements, organic and inorganic compounds in different phases of matter, and liquid water content (LWC), which is defined as the amount of condensed phase water in $g_{H2O}\ m^{-3}$ in different atmospheric media that include cloud (0.05–3), fog (0.1–0.3), adsorbed water in aerosols (10^{-5}–10^{-6}), and deliquesced aerosols (10^{-3}–10^{-2}) in equilibrium with gas-phase water.	[58, 59]
Oxidative potential	Ability of aerosols to catalytically form reactive oxygen species (ROS) such as hydroxyl radical ($^{\bullet}OH$), hydrogen peroxide (H_2O_2), hydroperoxyl radical ($HO_2{}^{\bullet}$), superoxide anion ($O_2{}^{\bullet-}$), peroxynitrites ($ONOO^-$), and organic peroxide (ROOR').	[60, 61]
	Chemical reactivity	
Interfacial or heterogeneous chemistry	Chemical and photochemical reactions taking place at interfaces: gas/liquid, gas/solid, liquid/solid, and in thin surface films.	[33, 62]
Multiphase chemistry	Chemical and photochemical reactions taking place among the three phases of matter: gas, liquid, and solid. They also include reactions in suspended micron-sized liquid droplets. Hence, bulk and heterogeneous processes are included in the definition.	[33, 39, 63]

1.2 Climate impacts

All of the aforementioned properties contribute to the direct and indirect effects of aerosol particles on the climate. The direct effect of aerosol particles refers to their role in modifying the planetary energy balance (i.e., radiative forcing) and precipitation, which has the highest uncertainty in climate and weather models [7]. Aerosol particles affect the radiative forcing directly through absorption and scattering of the incoming solar radiation. This effect is quantified through RFari, which is defined in Table 1.2 and shown in Figure 1.2 for different anthropogenic aerosol types, for the 1750–2010 period. The indirect effects of aerosols are related to cloud formation and lifetime and modification of atmospheric composition through heterogeneous and multiphase chemistry. Aerosol–cloud interactions are coupled by a multitude of dynamical and physical processes that span multitemporal and spatial scales (minutes to months, and meters to thousands of kilometers) [7]. Hence, aerosols not only contribute to changing atmospheric temperature but also the hydrological cycle and precipitation frequencies.

Figure 1.3 shows a schematic diagram of aerosol–cloud interactions from the fourth assessment report of the Intergovernmental Panel on Climate Change (IPCC) [64]. This diagram highlights aerosol–radiation interactions and the various radiative mechanisms associated with aerosol–cloud interactions and their effect on heating/cooling

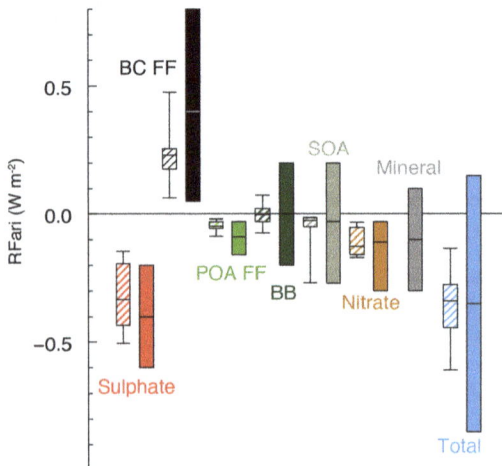

Figure 1.2: Annual mean top of the atmosphere radiative forcing due to aerosol–radiation interactions (RFari, in W m^{-2}) due to different anthropogenic aerosol types for the 1750–2010 period. Hatched whisker boxes show median (line), 5th–95th percentile ranges (box), and min/max values (whiskers) from AeroCom II models [65] corrected for the 1750–2010 period. Solid colored boxes show the AR5 best estimates and 90% uncertainty ranges. BC FF is for black carbon from fossil fuel and biofuel, POA FF is for primary organic aerosol from fossil fuel and biofuel, BB is for biomass burning aerosols, and SOA is for secondary organic aerosols. Figure and caption were reproduced from reference [2] with permission from Cambridge University Press, © 2013.

and precipitation. The size distribution of aerosols is a key property that impacts the extent of these effects and so are aerosol number, surface area, mass/volume, and composition (Figure 1.4) [30]. Primary aerosol particles listed in Table 1.1 are usually in the "coarse" size range, $d > 1$ µm. These particles dominate in terms of mass/volume (Figure 1.4), and hence, land biogeochemistry is preferentially affected by large particles and their composition [30]. Coarse mode particles influence the longwave (LW) radiation from the Earth (5–100 µm). The quantity of aerosol remote sensed via this radiation is referred to as LW aerosol optical depth (AOD). On the other hand, short wavelength AOD (SW AOD) in the 0.1–5 µm range from the Sun is heavily influenced by fine particles with $d < 1$ µm, which are mostly from anthropogenic activity, and their color and composition are of great importance. These fine mode particles are higher in number but contribute less mass compared to coarse mode particles. The surface area of these particles contributes to their atmospheric chemistry that alters their surface property. Anthropogenic aerosol particles cause up to ~60% increase in the number of cloud condensation nuclei. They also increase the number of ice nuclei particles [30].

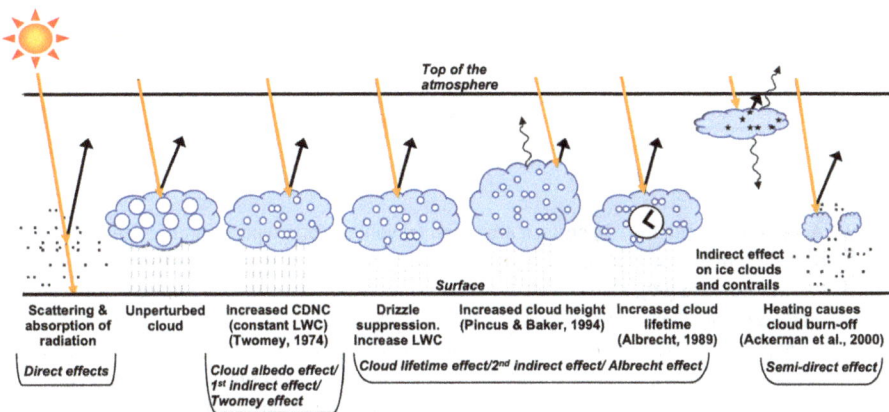

Figure 1.3: Cloud effects that have been identified as significant in relation to aerosols (modified from reference [66]). The small black dots represent aerosol particles; the larger open circles represent cloud droplets. Straight lines represent the incident and reflected solar radiation, and wavy lines represent terrestrial radiation. The filled white circles indicate cloud droplet number concentration (CDNC). The unperturbed cloud contains larger cloud drops as only natural aerosols are available as cloud condensation nuclei, while the perturbed cloud contains a greater number of smaller cloud drops as both natural and anthropogenic aerosols are available as cloud condensation nuclei (CCN). The vertical gray dashes represent rainfall, and LWC refers to the liquid water content. Figure and caption were reproduced from reference [64] with permission from Cambridge University Press, © 2007.

Eventually, aerosols get deposited on oceans, which cover nearly three-quarters of the Earth's surface [20]. Aerosols containing transition metals impact the biological

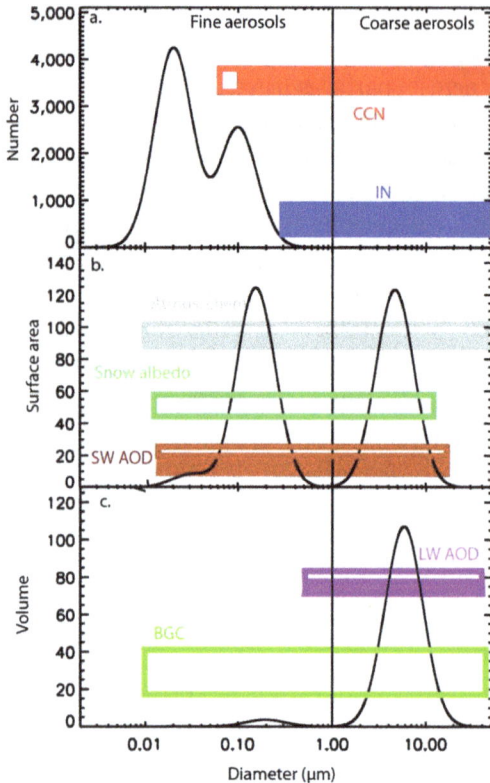

Figure 1.4: (a) Aerosol number, (b) surface area, and (c) volume for a typical trimodal aerosol distribution (from references [30, 67]) based on information in figure 7.6 in reference [68] and on information in reference [69]. Also shown in the boxes is a schematic representation of the typical aerosol diameter range impacting various processes as described in the text. Each process is assigned a panel depending on whether the impacts are primarily dependent on number (CCN and IN), surface area (SW AOD and LW AOD), or mass (biogeochemistry). Abbreviations: CCN, cloud condensation nuclei (red); IN, ice nuclei (blue); SW AOD, shortwave aerosol optical depth (brown); LW AOD, longwave aerosol optical depth (purple); and BGC, biogeochemically relevant species (green). Solid boxes represent only size-dependent processes, and the outlined boxes represent the part of the impact that is composition dependent. Figure and caption were reproduced with permission from reference [70] under the CC BY-NC-SA license, © 2013 The Authors.

productivity, carbon dioxide uptake [71], and atmospheric composition [72]. Climate models aim to quantify fluxes, formation, aging, transport, deposition, and biogeochemical processing of aerosols to better estimate their contribution to radiative forcing [39, 73–76]. Hence, the ultimate goal of field observations and laboratory measurements is providing data over a wide range of temporal and spatial scales, locations, and atmospheric conditions to validate and constrain aerosol chemistry and microphysics modules [23].

1.3 Chemical processing

Atmospheric aerosol particles provide unique multicomponent reaction environments whose reactivity can change the chemical composition of the gas and condensed phases. The term "atmospheric aging" refers to the processes that change the physicochemical properties of aerosols during their residence time in the atmosphere. These processes can take place at the surface of the particles or within the condensed phase. The physicochemical properties listed in Table 1.2 influence the rates of chemical processes during aging of aerosol particles. In general, the chemical and photochemical reactivities of atmospheric aerosol particles from primary and secondary sources can be grouped into two themes:
- Interfacial/heterogeneous reactions
- Multiphase reactions

The surface area-to-volume ratio of atmospheric aerosols determines the dominance of these reactions in changing the chemical composition and physical properties of the particles. This ratio changes with evaporation and water uptake processes due to changes in temperature and relative humidity.

Figure 1.5 shows a schematic diagram of some reactions and processes that highlight chemical coupling in the atmospheric gas, particle, and droplet phases. The majority of reactions with VOCs, CO, NO_x, and SO_2 in the gas phase are initiated and propagated by oxidants ($^\bullet OH$, O_3, H_2O_2, $HO_2^\bullet$, $RO_2^\bullet$, $NO_3^\bullet$, O_2, halogen radicals, etc.) [77–79] leading to the degradation of VOCs [80] or formation of SOA [13–15, 81]. These gases and their products could also adsorb or react on mineral and organic surfaces, depending on the amount of surface water (and the chemical composition of the surface) [82–85]. Hence, atmospheric aerosol particles and cloud/fog droplets can act as sinks or sources for atmospheric gases [83, 86, 87] and provide surfaces for heterogeneous reactions at the gas/liquid [88] or gas/solid (semisolid) interfaces [59, 83]. These atmospheric particles also act as seeds for condensation of low-volatility reaction products [15, 34, 53, 89]. Reaction products in cloud or surface water could be soluble [90, 91] or insoluble [92–94] in water. Hence, in multi-component systems containing organic and inorganic salts and water, partitioning between organic and aqueous phases can take place. Evaporation of liquid water content with decreasing relative humidity leads to efflorescence and liquid–liquid phase separation driven by the salting-out or salting-in effects [50, 95]. Therefore, atmospheric aging of particles changes the chemical composition of the gas and condensed phases, and can – in some cases – leads to particle growth through condensation and formation of clusters, oligomers, and polymers. Atmospheric aging of aerosol particles also leads to changing optical properties [47, 97], cloud condensation, and ice nucleation efficiency [36]. The following chapters provide specific examples and additional details.

Gas phase chemistry
- VOC oxidation/degradation forming OVOC ("high NOx")
- VOC oxidation and formation of ELVOC ("low NOx")
- Nitric acid formation from NOx
- Sulfuric acid formation from SO_2
- Reactions of sulfuric acid, nitric acid, methanesulfonic acid and organic acids with ammonia and amines
- Formation of OH, HO_2, and H_2O_2 from HOx cycle
- Formation of N_2O_5 and NO_3
- Halogen chemistry

Mineral

Organic/aqueous

Salts

Surface aqueous

Surface org

Partitioning/ release of gases

Evaporation

Water uptake

B

A

Interfacial chemistry (gas/solid or semi-solid)
- Ozonoloysis of organics
- Oxidation of organics by OH and other ROS/RNS
- Adsorption of inorganic gases and VOCs
- Surface reactions with oxides/clays/salts
- Bottom-up photogeneration of ROS
- Photoenhanced uptake of trace gases
- Catalysis of surface oligomerization/ polymerization

Interfacial chemistry (liquid/solid)
- Dissolution (proton, ligand, photo-driven)
- Complexation of organic and inorganic ligands
- Photogeneration of ROS such as H_2O_2
- Catalyzed surface oligomerization/ polymerization

Aqueous & Interfacial chemistry (gas/liquid)

$H_2O_2 + Fe(II) \rightarrow 2OH + Fe(III)$
$OH + RCHO \rightarrow RCO_2^- + H^+$
$SO_2 \cdot H_2O(aq) \rightarrow S(IV) + H^+$
$NH_3(aq) + H_2O \rightleftharpoons NH_4^+ + OH^-$
$S(IV) + TM + O_2 \rightarrow S(VI)$
$S(IV) + H_2O_2 \rightarrow S(VI)$
$S(IV) + O_3 \rightarrow S(VI)$
$HNO_3(aq) \rightarrow H^+ + NO_3^-$
$RCO_2^- + Fe(III) \rightarrow RCO_2Fe$
Polyphneols + Fe(III) → oligomers/polymers (s)
Diacids + Fe(III) → oligomers/polymers (s)
Photochem/aldol condes.

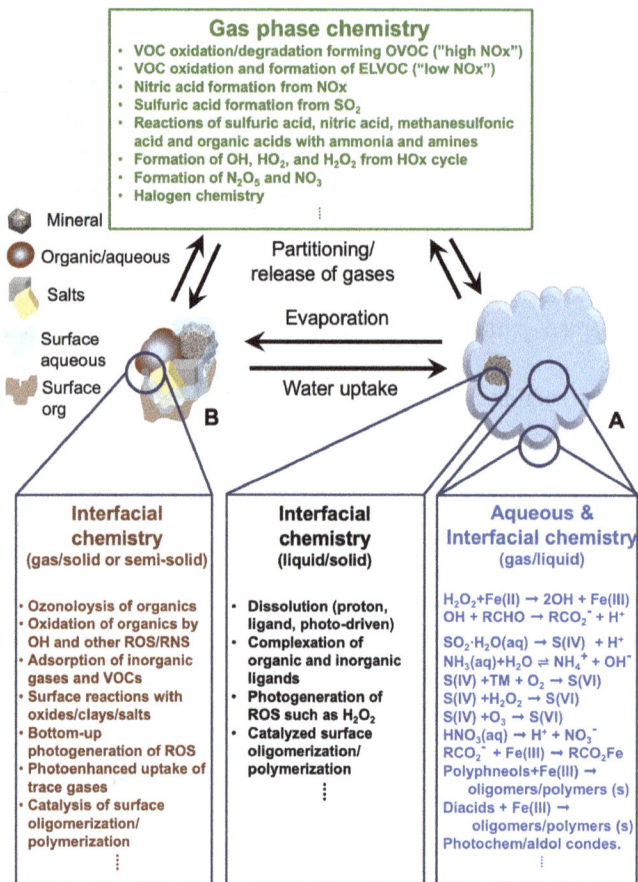

Figure 1.5: A schematic diagram of reactions and processes that highlight chemical coupling in the atmospheric gas, particle, and droplet phases. Day and night gas-phase chemistry leads to the degradation of VOCs, nucleation and growth of SOA, generation of reactive radicals, and transformation of NO_x and SO_2. Gas-phase reactants and products can partition to cloud/ fog droplet (A) or aerosol particles (B). The organic content in (B) could be from primary or secondary sources. A cloud/fog droplet (A) is a microreactor for bulk and heterogeneous chemistry at the liquid/solid or semisolid and gas/liquid interfaces. Evaporation processes decrease the amount of liquid water in aerosol particles (B), leading to crystallization of salts and preferential "salting out" of organics. Reactions in (A) and (B) can release gases as well. Abbreviations: OVOC, oxygenated volatile organic compounds; ELVOC, extremely low VOC; and TM, transition metals. Reproduced with permission from reference [96] under the CC-BY-NC license, © 2021 The Author(s).

References

[1] Calvert JG. Glossary of atmospheric chemistry terms. Pure and Appl Chem. 1990;62(11): 2167–2219.

[2] Boucher O, Randall D, Artaxo P, Bretherton C, Feingold G, Forster F, et al. Clouds and Aerosols. In: Stocker TF, Qin D, Plattner G-K, Tignor M, Allen SK, Boschung J, et al. editors. Climate Change 2013: The Physical Science Basis Contribution of Working Group I to the Fifth Assessment Report of the Intergovernmental Panel on Climate Change. Cambridge, United Kingdom and New York, NY, USA: Cambridge University Press; 2013. p. 571–657.

[3] Heintzenberg J, Raes F, Schwartz SE. Tropospheric Aerosols. In: Brasseur G, Prinn RG, Pszenny AAP, editors. Atmospheric Chemistry in a Changing World – an Integration and Synthesis of a Decade of Tropospheric Chemistry Research. Berlin: Springer; 2003. p. 125–156.

[4] Finlayson-Pitts BJ, Pitts JN Jr. Chemistry of the Upper and Lower Atmosphere. New York: Academic Press; 2000.

[5] Seinfeld JH, Pandis SN. Atmospheric Chemistry and Physics: From Air Pollution to Climate Change. New York: Wiley; 2006.

[6] Tomasi C, Lupi A. Primary and Secondary Sources of Atmospheric Aerosol. In: Tomasi C, Fuzzi S, Kokhanovsky A, editors. Atmospheric Aerosols: Life Cycles and Effects on Air Quality and Climate. Weinheim, Germany: Wiley-VCH Verlag GmbH & Co.; 2016. p. 1–86.

[7] Boucher O. Atmospheric Aerosols: Properties and Impacts. France: Springer; 2015.

[8] Kerminen V-M, Chen X, Vakkari V, Petaja T, Kulmala M, Bianchi F. Atmospheric new particle formation and growth: Review of field observations. Environ Res Lett. 2018;13(103003):1–38.

[9] Philip S, Martin RV, Snider G, Weagle CL, van Donkelaar A, Brauer M, et al. Anthropogenic fugitive, combustion and industrial dust is a significant, underrepresented fine particulate matter source in global atmospheric models. Environ Res Lett. 2017;12(044018):1–7.

[10] Weichenthal SA, Pollitt KG, Villeneuve PJ. PM2.5, oxidant defence and cardiorespiratory health: A review. Environ Health. 2013;12(40):1–8.

[11] Shi W, Li T, Zhang Y, Sun Q, Chen C, Wang J, et al. Depression and anxiety associated with exposure to fine particulate matter constituents: A cross-sectional study in North China. Environ Sci Technol. 2020;54(24):16006–16016.

[12] Landrigan PJ, Fuller R, Acosta NJR, Adeyi O, Arnold R, Basu N, et al. The Lancet Commission on pollution and health. Lancet. 2018;391:462–512.

[13] Bruggemann M, Xu R, Tilgner A, Kwon KC, Mutzel A, Poon HY, et al. Organosulfates in ambient aerosol: State of knowledge and future research directions on formation, abundance, fate, and importance. Environ Sci Technol. 2020;54:3767–3782.

[14] Lim YB, Kim H, Kim JY, Turpin BJ. Photochemical organonitrate formation in wet aerosols. Atmos Chem Phys. 2016;16:12631–12647.

[15] Shrivastava MK, Cappa CD, Fan J, Goldstein AH, Guenther AB, Jimemez JL, et al. Recent advances in understanding secondary organic aerosol: Implications for global climate forcing. Rev Geophys. 2017;55:509–559, doi:10.1002/2016RG000540.

[16] Hallquist M, Wenger JC, Baltensperger U, Rudich Y, Simpson D, Claeys M, et al. The formation, properties and impact of secondary organic aerosol: Current and emerging issues. Atmos Chem Phys. 2009;9:5155–5236.

[17] Steiner AL. Role of the terrestrial biosphere in atmospheric chemistry and climate. Acc Chem Res. 2020;53(7):1260–1268.

[18] Hoesly RM, Smith SJ, Feng L, Klimont Z, Janssens-Maenhout G, Pitkanen T, et al. Historical (1750–2014) anthropogenic emissions of reactive gases and aerosols from the community emissions data system (CEDS). Geosci Model Dev. 2018;11:369–408.

[19]   Jacob DJ. Introduction to Atmospheric Chemistry. Princeton, N.J.: Princeton University Press; 1999.

[20]   Kanakidou M, Myriokefalitakis S, Tsigaridis K. Aerosols in atmospheric chemistry and biogeochemical cycles of nutrients. Environ Res Lett. 2018;13(1–22):063004.

[21]   Mahowald N. Aerosol indirect effect on biogeochemical cycles and climate. Science. 2011;334:794–796.

[22]   Scanza RA, Mahowald N, Ghan S, Zender CS, Kok JF, Liu X, et al. Modeling dust as component minerals in the community atmosphere model: Development of framework and impact on radiative forcing. Atmos Chem Phys. 2015;15:537–561.

[23]   Textor C, Schulz M, Guibert S, Kinne S, Balkanski Y, Bauer S, et al. Analysis and quantification of the diversities of aerosol life cycles within Aerocom. Atmos Chem Phys. 2006;6:1777–1813.

[24]   Chen H, Grassian VH. Iron dissolution of dust source materials during simulated acidic processing: The effect of sulfuric, acetic, and oxalic acids. Environ Sci Technol. 2013;47(18): 10312–10321.

[25]   Ito A. Atmospheric processing of combustion aerosols as a source of bioavailable iron. Environ Sci Technol Lett. 2015;2(3):70–75.

[26]   Lighty JS, Vernath JM, Sarofim AF. Combustion aerosols: Factors governing their size and composition and implications to human health. J Air Waste Manage Assoc. 2000;50: 1565–1618.

[27]   Estillore AD, Trueblood JV, Grassian VH. Atmospheric chemistry of bioaerosols: Heterogeneous and multiphase reactions with atmospheric oxidants and other trace gases. Chem Sci. 2016;7:6604–6616.

[28]   Bertram TH, Cochran RE, Grassian VH, Stone EA. Sea spray aerosol chemical composition: Elemental and molecular mimics for laboratory studies of heterogeneous and multiphase reactions. Chem Soc Rev. 2018;47:2374–2400.

[29]   Quinn PK, Collins DB, Grassian VH, Prather KA, Bates TS. Chemistry and related properties of freshly emitted sea spray aerosol. Chem Rev. 2015;115:4383–4399.

[30]   Mahowald N, Ward DS, Kloster S, Flanner MG, Heald CL, Heavens NG, et al. Aerosol impacts on climate and biogeochemistry. Ann Rev Environ Resour. 2011;36:45–74.

[31]   Gentner DR, Jathar SH, Gordon TD, Bahreini R, Day DA, El Haddad I, et al. Review of urban secondary organic aerosol formation from gasoline and diesel motor vehicle emissions. Environ Sci Technol. 2017;51:1074–1093.

[32]   Wennberg PO, Bates KH, Crounse JD, Dodson LG, McVay RC, Mertens LA, et al. Gas-phase reactions of isoprene and its major oxidation products. Chem Rev. 2018;118:3337–3390.

[33]   Rao G, Vejerano EP. Partitioning of volatile organic compounds to aerosols: A review. Chemosphere. 2018;212:282–296.

[34]   Kroll JH, Seinfeld JH. Chemistry of secondary organic aerosol: Formation and evolution of low-volatility organics in the atmosphere. Atmos Environ. 2008;42:3593–3624.

[35]   Kerminen V-M, Paramonov M, Anttila T, Riipinen I, Fountoukis C, Korhonen H, et al. Cloud condensation nuclei production associated with atmospheric nucleation: A synthesis based on existing literature and new results. Atmos Chem Phys. 2012;12:12037–12059.

[36]   Tang M, Cziczo DJ, Grassian VH. Interactions of water with mineral dust aerosol: water adsorption, hygroscopicity, cloud condensation, and ice nucleation. Chem Rev. 2016;116: 4205–4259.

[37]   Knopf DA, Alpert PA, Wang B. The role of organic aerosol in atmospheric ice nucleation: A review. ACS Earth Space Chem. 2018;2:168–202.

[38]   Kanji ZA, Ladino LA, Wex H, Boose Y, Burkert-Kohn M, Cziczo DJ, et al. Overview of ice nucleating particles. Meteorol Monogr. 2017;58:1–33.

[39] Stevens R, Dastoor A. A review of the representation of aerosol mixing state in atmospheric models. Atmosphere. 2019;10(168), doi:10.3390/atmos10040168.

[40] Riemer N, Ault AP, West M, Craig RL, Curtis JH. Aerosol mixing state: Measurements, modeling, and impacts. Rev Geophys. 2019;57(2):187–249.

[41] Ault AP, Axson JL. Atmospheric aerosol chemistry: Spectroscopic and microscopic advances. Anal Chem. 2017;89:430–452.

[42] Riemer N, West M. Quantifying aerosol mixing state with entropy and diversity measures. Atmos Chem Phys. 2013;18:12595–12612.

[43] China S, Mazzoleni C. Preface: Morphology and internal mixing of atmospheric particles. Atmosphere. 2018;9(249):1–6.

[44] Gorkowski K, Donahue NM, Sullivan RC. Aerosol optical tweezers constrain the morphology evolution of liquid-liquid phase-separated atmospheric particles. Chem. 2020;6:204–220.

[45] Song M, Marcolli C, Krieger UK, Lienhard DM, Peter T. Morphologies of mixed organic/inorganic/aqueous aerosol droplets. Faraday Discuss. 2013;165(1):289–316.

[46] Ravishankara AR, Rudich Y, Wuebbles DJ. Physical chemistry of climate metrics. Chem Rev. 2015;115:3682–3703.

[47] Laskin A, Laskin J, Nizkorodov SA. Chemistry of atmospheric brown carbon. Chem Rev. 2015;115(10):4335–4382.

[48] Ziemann PJ. Phase matters for aerosols. Nature. 2010;467:797–798.

[49] Grover PK, Ryall RL. Critical appraisal of salting-out and its implications for chemical and biological sciences. Chem Rev. 2005;105(1):1–10.

[50] You Y, Smith ML, Song M, Martin ST, Bertram AK. Liquid–liquid phase separation in atmospherically relevant particles consisting of organic species and inorganic salts. Int Rev Phys Chem. 2014;33(1):43–77.

[51] Fundamentals of viscosity. [Available from: https://resources.saylor.org/wwwresources/archived/site/wp-content/uploads/2011/04/Viscosity.pdf.

[52] Koop T, Bookhold J, Shiraiwa M, Pöschl U. Glass transition and phase state of organic compounds: Dependency on molecular properties and implications for secondary organic aerosols in the atmosphere. Phys Chem Chem Phys. 2011;13(43):19238–19255.

[53] Donahue NM, Robinson AL, Trump ER, Riipinen I, Kroll JH. Volatility and Aging of Atmospheric Organic Aerosol. In: McNeill VF, Ariya P, editors. Atmospheric and Aerosol Chemistry Topics in Current Chemistry, vol. 339, Berlin, Heidelberg: Springer; 2012. p. 97–143.

[54] Robinson AL, Donahue NM, Shrivastava MK, Weitkamp EA, Sage AM, Grieshop AP, et al. Rethinking organic aerosols: semivolatile emissions and photochemical aging. Science. 2007;315(5816):1259–1262.

[55] Craig RL, Ault AP. Aerosol Acidity: Direct Measurement from a Spectroscopic Method. In: Hunt SW, Laskin A, Nizkorodov SA, editors. Multiphase Environmental Chemistry in the Atmosphere. Washington DC: ACS; 2018. p. 171–191.

[56] Weber RJ, Guo H, Russell AG, Nenes A. High aerosol acidity despite declining atmospheric sulfate concentrations over the last 15 years. Nature Geosci. 2016;9:282–285.

[57] Pye HOT, Nenes A, Alexander B, Ault AP, Barth MC, Clegg SL, et al. The acidity of atmospheric particles and clouds. Atmos Chem Phys. 2020;20:4809–4888.

[58] McNeill VF. Aqueous organic chemistry in the atmosphere: Sources and chemical processing of organic aerosols. Environ Sci Technol. 2015;49:1237–1244.

[59] Zhao R, Lee AKY, Wang C, Wania F, Wong JPS, Zhou S, et al. The Role of Water in Organic Aerosol Multiphase Chemistry: Focus on Partitioning and Reactivity. In: Barker JR, Steiner AL, Wallington TJ, editors. Advances in Atmospheric Chemistry. Vol. 2, New Jersey: World Scientific; 2019. p. 95–184.

[60] Charrier JG, Anastasio C. On dithiothreitol (DTT) as a measure of oxidative potential for ambient particles: Evidence for the importance of soluble transition metals. Atmos Chem Phys. 2012;12:9321–9333.

[61] Pöschl U, Shiraiwa M. Multiphase chemistry at the atmosphere–biosphere interface influencing climate and public health in the anthropocene. Chem Rev. 2015;115(10): 4440–4475.

[62] Pillar EA, Guzman MI. An overview of dynamic heterogeneous oxidations in the troposphere. Environments. 2018;5(9), 104, doi:10.3390/environments5090104.

[63] Bianco A, Passananti M, Brigante M, Mailhot G. Photochemistry of the cloud aqueous phase: A review. Molecules. 2020;25(2):423, doi:10.3390/molecules25020423.

[64] Forster P, Ramaswamy V, Artaxo P, Berntsen T, Betts R, Fahey DW, et al. Changes in atmospheric constituents and in radiative forcing. In: Solomon S, Qin D, Manning M, Chen Z, Marquis M, Averyt KB, et al. editors. Climate Change 2007: The Physical Science Basis Contribution of Working Group I to the Fourth Assessment Report of the Intergovernmental Panel on Climate Change. Cambridge, United Kingdom and New York, N.Y.: Cambridge University Press; 2007. p. 130–234.

[65] Myhre G, Samset BH, Schulz M, Balkanski Y, Bauer S, Berntsen TK, et al. Radiative forcing of the direct aerosol effect from Aerocom phase II simulations. Atmos Chem Phys. 2013;13: 1853–1877.

[66] Haywood JM, Boucher O. Estimates of the direct and indirect radiative forcing due to tropospheric aerosols: A review. Rev Geophys. 2000;38(4), 513–543.

[67] Mahowald N, Muhs D, Levis S, Rasch P, Yoshioka M, Zender CS. Change in atmospheric mineral aerosols in response to climate: Last glacial period, pre-industrial, modern and double-carbon dioxide climates. J Geophys Res. 2006;111:D10202 (doi: 10.1029/2005JD006653).

[68] Seinfeld JH, Pandis SN. Atmospheric Chemistry and Physics: From Air Pollution to Climate Change. New York: Wiley; 1998.

[69] Dulac F, Moulin C, Lambert CE, Guillard F, Poitou J, Guelle W, et al. Qualitative remote sensing of African dust transport to the Mediterranean. In: Guerzoni S, Chester R, editors. The Impact of Desert Dust across the Mediterranean. Netherlands: Kluwer Academic Publishers; 1996. p. 25–49.

[70] Mahowald N, Albani S, Kok JF, Engelstaedter S, Scanza RA, Ward DS, et al. The size distribution of desert dust aerosols and its impact on the Earth system. Aeolian Res. 2014;15: 53–71.

[71] Meskhidze N, Volker C, Al-Abadleh HA, Barbeau K, Bressac M, Buck C, et al. Perspective on identifying and characterizing the processes controlling iron speciation and residence time at the atmosphere-ocean interface. Mar Chem. 2019;217(1–16):103704.

[72] Ryder OS, Campbell NR, Shaloski M, Al-Mashat H, Nathanson GM. Role of organics in regulating ClNO2 production at the air–sea interface. J Phys Chem A. 2015;119(31): 8519–8526.

[73] Semeniuk K, Dastoor A. Current state of atmospheric aerosol thermodynamics and mass transfer modeling: A review. Atmosphere. 2020;11(156), doi:10.3390/atmos11020156.

[74] Ervens B. Modeling the processing of aerosol and trace gases in clouds and fogs. Chem Rev. 2015;115(10):4157–4198.

[75] Kanakidou M, Seinfeld JH, Pandis SN, Barnes I, Dentener FJ, Facchini MC, et al. Organic aerosol and global climate modelling: A review. Atmos Chem Phys. 2005;5:1053–1123.

[76] Ito A. Global modeling study of potentially bioavailable iron input from shipboard aerosol sources to the ocean. Global Biogeochem Cyc. 2013;27:1–10.

[77]   Monks PS. Gas-phase radical chemistry in the troposphere. Chem Soc Rev. 2005;34: 376–395.

[78]   Brown SS, Ryerson TB, Wollny AG, Brock CA, Peltier R, Sullivan AP, et al. Variability in nocturnal nitrogen oxide processing and its role in regional air quality. Science. 2006;311: 67–80.

[79]   Kleffmann J. Daytime sources of nitrous acid (HONO) in the atmospheric boundary layer. Chem Phys Chem (Mini Review). 2007;8:1137–1144.

[80]   Atkinson R, Arey J. Atmospheric degradation of volatile organic compounds. Chem Rev. 2003;103:4605.

[81]   Chen H, Varner ME, Gerber B, Finlayson-Pitts BJ. Reactions of methanesulfonic acid with amines and ammonia as a source of new particles in air. J Phys Chem B. 2016;120:1526–1536.

[82]   Cwiertny DM, Young MA, Grassian VH. Chemistry and photochemistry of mineral dust aerosol. Annu Rev Phys Chem. 2008;59:27–51.

[83]   George C, Ammann M, D'Anna B, Donaldson DJ, Nizkorodov SA. Heterogeneous photochemistry in the atmosphere. Chem Rev. 2015;115(10):4218–4258.

[84]   Donaldson DJ, Valsaraj KT. Adsorption and reaction of trace gas-phase organic compounds on atmospheric water film surfaces: A critical review. Environ Sci Technol. 2010;44:865–873.

[85]   George IJ, Abbatt JPD. Heterogeneous oxidation of atmospheric aerosol particles by gas-phase radicals. Nature Chem. 2010;2:713–722.

[86]   Harris E, Sinha B, van Pinxteren D, Tilgner A, Fomba KW, Schneider J, et al. Enhanced role of transition metal ion catalysis during in-cloud oxidation of SO2. Science. 2013;340:727–730.

[87]   Chen Q, Sherwen T, Evans M, Alexander B. DMS oxidation and sulfur aerosol formation in the marine troposphere: A focus on reactive halogen and multiphase chemistry. Atmos Chem Phys. 2018;18:13617–13637.

[88]   Donaldson DJ, Vaida V. The influence of organic films at the air-aqueous boundary on atmospheric processes. Chem Rev. 2006;103(12):4717–4729.

[89]   Murphy SM, Sorooshian A, Kroll JH, Ng NL, Chhabra PS, Tong C, et al. Secondary aerosol formation from atmospheric reactions of aliphatic amines. Atmos Chem Phys. 2007;7: 2313–2337.

[90]   Herrmann H, Schaefer T, Tilgner A, Styler SA, Weller C, Teich M, et al. Tropospheric aqueous-phase chemistry: Kinetics, mechanisms, and its coupling to a changing gas phase. Chem Rev. 2015;115(10):4259–4334.

[91]   Hoyle CR, Fuchs C, Järvinen E, Saathoff H, Dias A, El Haddad I, et al. Aqueous phase oxidation of sulphur dioxide by ozone in cloud droplets. Atmos Chem Phys. 2016;16:1693–1712.

[92]   Al Nimer A, Rocha L, Rahman MA, Nizkorodov SA, Al-Abadleh HA. Effect of oxalate and sulfate on iron-catalyzed secondary brown carbon formation. Environ Sci Technol. 2019;53(12): 6708–6717.

[93]   Tran A, William G, Younus S, Ali NN, Blair SL, Nizkorodov SA, et al. Efficient formation of light-absorbing polymeric nanoparticles from the reaction of soluble Fe(III) with C4 and C6 dicarboxylic acids. Environ Sci Technol. 2017;51(17):9700–9708.

[94]   Slikboer S, Grandy L, Blair SL, Nizkorodov SA, Smith RW, Al-Abadleh HA. Formation of light absorbing soluble secondary organics and insoluble polymeric particles from the dark reaction of catechol and guaiacol with Fe(III). Environ Sci Technol. 2015;49(13):7793–7801.

[95]   Wang C, Lei YD, Endo S, Wania F. Measuring and modeling the salting-out effect in ammonium sulfate solutions. Environ Sci Technol. 2014;48(22):13238–13245.

[96]   Al-Abadleh HA. Aging of atmospheric aerosols and the role of iron in catalyzing brown carbon formation. Environ Sci: Atmos. 2021, 1(6), 291–474.

[97]   Seinfeld JH. Black carbon and brown clouds. Nature Geosci. 2008;1:15–16.

Chapter 2
Instrumentation for measuring aerosols' physical and chemical properties

Quantifying that impact of atmospheric aerosols on the climate requires measuring their number and properties with time. McMurry published a review of atmospheric aerosol measurements of physical and chemical properties [1], which are classified into categories according to the instrumental capacity to resolve size, time, and composition as shown in Figure 2.1. A concise summary of instruments that measure total (i.e., integral) or size-resolved physical and chemical properties in McMurry's review [1] is provided in Table 2.1. Since McMurry's paper, edited books [2, 3] and a number of reviews published on advanced analytical tools [4] were used to study hygroscopic properties and water uptake [5], ice nucleation (IN) [6, 7], aerosol morphology and mixing states [8, 9], optical properties [10], viscosity [11], liquid–liquid-phase separation [12], acidity [13, 14], and chemical composition [15–20]. In this chapter, a synthesis of relevant literature is provided with highlights to recent developments in aerosol analytical tools.

https://doi.org/10.1515/9781501519376-002

Instrument	Measured Quantity and Resolution

Perfect Single Particle Measurements of Size-Resolved Composition

Measured Quantity $= N_\infty \cdot gdvdn_l$

Distribution Function / Size (Continuous) / Time (Continuous) / Composition (Complete)

Measurements of Integral or Size-Resolved Physical Properties

"Continuous" measurements of integral properties

CNC
CCN
mass concentration
Epiphaniometer
integrating nephelometer
photoacoustic spectrometer

Measured Quantity $= N_\infty \cdot \int_0^\infty W(v)g(t)dvdn_l$

Distribution Function / Time: Seconds / Composition

Time-integrated measurements of size-resolved mass:

Cascade Impactor

Measured Quantity $= N_\infty \cdot \int\int_{v_1}^{v_2} \rho_p \cdot v \cdot gdn_l dv$

Distribution Function / Size / Time: Hours / Composition

Physical size distributions:

optical particle counter
electrical mobility analyzer
aerodynamic particle sizer
diffusion battery

Measured Quantity $= N_\infty \cdot \int_{v_1}^{v_2} gdvdn_l$

Distribution Function / Size / Time: -Minutes / Composition

Measurements of Integral or Size-Resolved Chemical Properties

Time and size integrated measurements of composition:

Filter

Measured Quantity $= N_\infty \cdot \int\int_0^\infty n_l \cdot gdn_l dv$

Distribution Function / Size / Time: Hours / Composition

Time-integrated measurements of size-resolved composition:

Cascade Impactor
Electron Microscopy
Laser microprobe

Measured Quantity $= N_\infty \cdot \int\int_{v_1}^{v_2} gn_l dvdn_l$

Distribution Function / Size / Time: Hours / Composition

Real-time measurement of individual particle composition

Mass spectrometer for individual particle analysis

Measured Quantity $= N_\infty \cdot gdvdn_l$

Distribution Function / Time: -Seconds / Composition / Not Detected

---

OPC = Optical Particle Counter
SEMS = Scanning Electrical Mobility Spectrometer
SMPS = Scanning Mobility Particle Spectrometer
CNC = Condensation Nucleus (or Nuclei) Counter
CPC = Condensation Particle Counter (the same as a CNC)

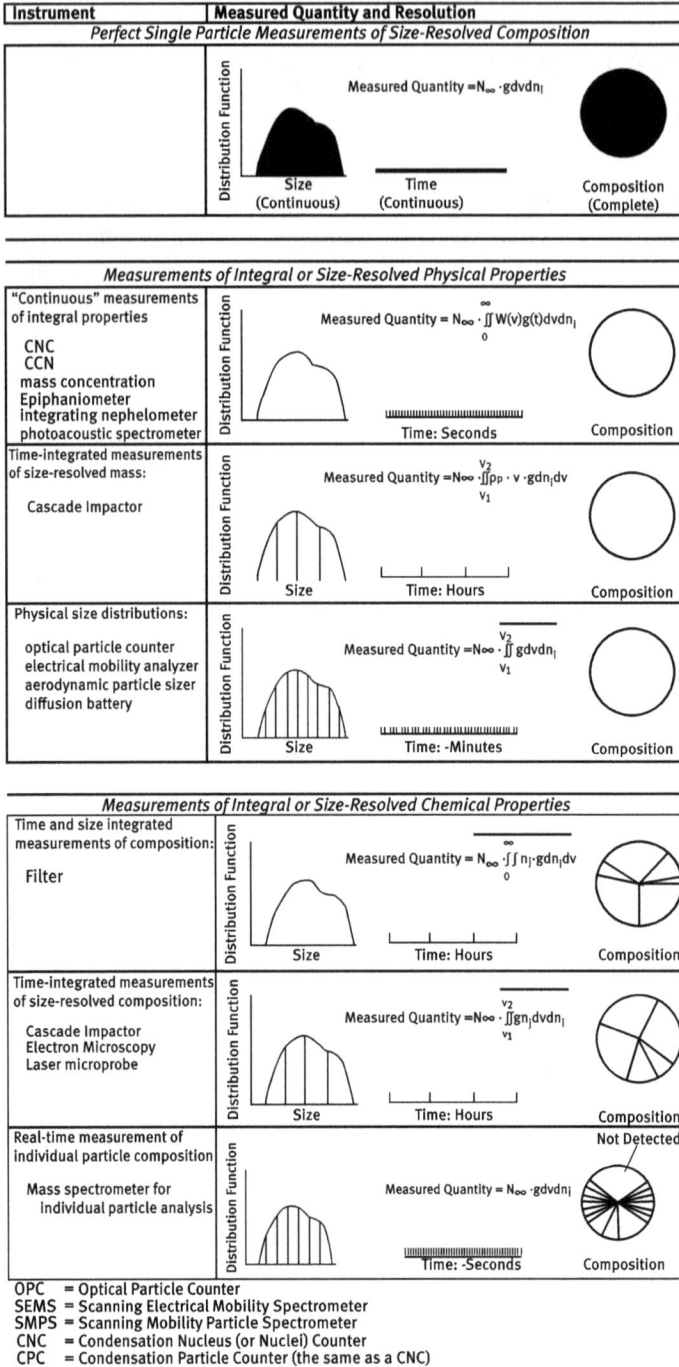

Figure 2.1: Classification of aerosol instruments according to their capacity to resolve size, time, and composition. Figure and caption are reprinted from reference [1] with permission, © 2020 Elsevier.

Table 2.1: A summary of atmospheric aerosol measurements by McMurry's review [1].

Measurement type	Instrumentation	Summary
	Physical properties of aerosol	
a) Integral		Provides integrals of specified variables over a given size range (Figure 2.1)
a1) Number concentration	Condensation nucleus counters (CNCs); condensation particle counters (CPCs); Aitken nuclei counters (ANCs)	Measure the total aerosol number concentration larger than some minimum detectable size (3 nm); direct single-particle counting or optical detection of particles formed by condensation with diameter growth factors 100–1,000 using supersaturated vapor of water or n-butyl alcohol
a2) CCN concentration	CCN counters	Measure concentration of particles (min. $d = 40$ nm) converted to cloud droplets by water condensation (i.e., activation) at a specified supersaturation (0.01–1%). The saturation ratio that is required to activate particles increases with decreasing size; uses thermal gradient diffusion chambers to produce the desired supersaturations.
a3) Particle mass concentrations	Filtration	*Manual methods:* Using filter samplers with inlets that eliminate particles above a specified size cut; increase in filter mass for a known volume of air sampled; particles <100 nm are collected by diffusion (collection efficiency ↑ as size ↓); particles >500 nm are collected by interception and impaction (collection efficiency ↑ as size ↑). *Automated methods:* Beta gauge, piezoelectric crystals, and the oscillating element instruments.

Table 2.1 (continued)

Measurement type	Instrumentation	Summary
a4) Epiphaniometer		Measures the diffusion-limited mass transfer rate of a gas to aerosol particles. These measurements provide information on the maximum possible rates of vapor condensation or gas reaction with the aerosol.
a5) Optical properties	Integrating nephelometers	*Scattering coefficient:* A measure of the total amount of light scattered by an aerosol; obeys the same power-law relationships as aerosol size distribution function. Hence, size distribution is wavelength dependent.
	Photoacoustic spectroscopy	*Absorption coefficient:* Filter technique (measures light transmittance through filter; affected by scattering and absorption; does not necessarily reflect optical properties of airborne particles). Photoacoustic spectroscopy (measures absorption of suspended particles in real time). Elemental carbon (EC) concentrations (used to infer aerosol absorption coefficient of particles on filters or airborne; require accurate measurement of the EC concentration and their "mass absorption efficiency").

Table 2.1 (continued)

Measurement type	Instrumentation	Summary
b) Size-resolved		
	b1) Optical particle counters	Measure the intensity of light (laser or white) scattered by individual particles as they traverse a tightly focused beam of light; need a heat exchanger to minimize error caused by aerosol heating that affects the size; angular distribution of scattered light (i.e., differential light scattering) in conjunction with the Mie theory can be used to infer refractive indices and particle shape.
	b2) Aerodynamic particle size	Measure particle velocity from an aerosol rapidly accelerated through a nozzle to determine size, which is closely related to aerodynamic diameter, defined as the diameter of a unit density sphere that has the same settling velocity as the particle; scattered light from particles traveling a known distance is used to detect the particles; sizing errors stem from change in relative humidity and deformation in shapes of liquid droplets due to flow cooling and pressure drop associated with expansion through the nozzle.

Table 2.1 (continued)

Measurement type	Instrumentation	Summary
	b3) Differential mobility particle sizer (DMPS); scanning mobility particle spectrometer (SMPS)	Includes a differential mobility analyzer (DMA) and a particle detector (a CNC or aerosol electrometer); measures diameter range 3–500 nm; particles in the aerosol are exposed to a bipolar cloud of ions to achieve the Boltzmann charge equilibrium; classifies particles according to electrical mobility; clearly defined for spherical particles; primary limitations are detecting very low concentrations of very small particles and multiple charging.
	b4) Diffusion batteries	Measure aerosol number concentration downstream of each collecting element through which aerosol flows (e.g., fine wire mesh screens and series of fine capillaries); they are based on the principle that particle diffusivity increases with decreasing size; hence, the rate at which they deposit on nearby surfaces increases, and the decay in aerosol concentration through the series of collecting elements is related to size distribution.
c) Aerosol water content	Gravimetric measurements using microbalance; tandem differential mobility analyzer (TDMA)	Microbalance measures mass before and after water uptake with increasing relative humidity (RH); in TDMA, growth factors are calculated as a function of RH from the dependence of size on RH.
d) Aerosol volatility	Indirect measurements using two CNCs	Measure the effect of volatilization on number concentrations and size distributions

Table 2.1 (continued)

Measurement type	Instrumentation	Summary
e) Particle density	DMA-impactor technique for particles of known composition, size 0.06–0.18 µm	DMA measures electromobility equivalent size (density independent); impactor measures particles' aerodynamic size (density dependent).
	Chemical composition of aerosol	
a) Off-line measurements		Sampling artifacts include volatilization of semivolatile compounds, changes in temperature, RH, composition during sampling, transport and storage, use of diffusion denuder, and after-Teflon filters help in quantifying evaporative losses.
	a1) Filter sampling	Analysis of deposits on filter substrates in filter samplers
	a2) Impactors	Classify particles according to their aerodynamic diameter, that is, particle density dependent using a series of carbon-free stages, each with a successfully smaller cut point (in cascade impactors); coatings or sampling at high RH needed to eliminate particle bounce. Chemical analysis follows to get size-dependent chemical composition.
	a3) Laser microprobe MS	Off-line analysis of particles collected on a substrate, which are irradiated with a high-power pulse laser, and the ejected ion fragments are analyzed by mass spectrometry; chemical composition is altered due to vacuum, chemical reactions, and evaporation.

Table 2.1 (continued)

Measurement type	Instrumentation	Summary
	a4) Electron microscopy	Particle morphology and elemental composition for individual particles on a filter or impaction substrate; standardization needed to get quantitative elemental analysis; volatilization can occur due to vacuum and heating by electron beam; environmental scanning electron microscope.
b) Real-time measurements		
	b1) Particulate carbon analyzers; ambient carbon particle monitor	Samples are collected on filters followed by heating to volatilize organic carbon in helium at high temperature to convert to CO_2 and water and then to methane.
	b2) Sulfur and nitrogen species analyses	Flame photometric detectors for detecting S_2 formed when sulfur compounds are burned in a hydrogen-rich fame; for particulate nitrate, a chemiluminescent NO_x analyzer is used after flash vaporization of particles collected on a single-stage impactor; ion chromatography.
	b3) Single-particle mass spectrometry	Real-time in situ techniques for the analysis of individual particles in a flowing gas stream; rapid depressurization of the aerosol, formation of a particle beam, and irradiation of particles by a high-power pulse laser to produce ions that are analyzed by rapid depressurization of the aerosol, formation of a particle beam, and irradiation of particles by a high-power pulse laser to produce ions that are analyzed by mass spectrometry.

2.1 Measuring physical properties

2.1.1 Hygroscopic properties and ice nucleation

There are two recent reviews that summarized the advantages and limitations of major techniques used in aerosol hygroscopicity studies (see Table 2.2 from reference [5]) and those utilized for single-particle measurements [21]. These techniques can be grouped according to the parameter they measure (i.e., working principle): (a) water vapor pressure change, (b) sample mass of deposited or levitated particles, (c) particle morphology, (d) particle diameter, (e) light extinction, (f) binding energy of oxygen atoms in the lattice versus surface hydroxyl groups and water, and (g) light scattering properties. Spectroscopic techniques such as Raman and infrared spectroscopy and nonlinear optical techniques such as sum-frequency generation (SFG) have also been used where spectra are collected as a function of relative humidity (RH) for particles deposited on substrates [5]. The spectral features assigned to water vibrational modes in the condensed phase are sensitive to the amount of water and molecular environment reflecting the heterogeneity of the hydrogen bonding network. Hygroscopicity studies conducted using infrared spectroscopy yield relative amounts of water. The analysis of the stretching region of hydroxyl groups showed structural differences between bulk and interfacial water, which has impacts on the heterogeneous chemistry of atmospheric aerosol particles [22–26]. The use of atmospheric pressure X-ray photoelectron spectroscopy (XPS) for studying water uptake on flat single-crystal model oxide surfaces provided mechanistic information on the binding mechanism [27]. In Chapter 3, specific examples are presented on the hygroscopic properties of field-collected and lab proxies of atmospheric aerosol particles.

Table 2.2: Summary and comparison of key features of major techniques for aerosol hygroscopicity measurements.

	Isopiestic method	Nonisopiestic method	Physisorption analyzer	Katharometer
(1) Working principle	Measures water vapor pressure of a solution	Measures water vapor pressure of a solution	Measures water vapor change when exposure to particles	Measures water vapor change when exposure to particles
(2) Sample status	Bulk solution	Bulk solution	Particles deposited on substrates	Particles deposited on substrates
(3) Size range	Not applicable	Not applicable	Not applicable	Not applicable

Table 2.2 (continued)

(4) Supersaturated samples	No	No	No	Yes
(5) Nonspherical particles	Yes	Yes	Yes	Yes
(6) Water adsorption	No	No	Yes	No
(7) Ambient application	No	No	Yes (off-line)	Yes (off-line)
	Analytical balance	**Thermogravimetric analysis**	**QCM**	**Optical microscopy**
(1) Working principle	Measures sample mass at different RH	Measures sample mass at different RH	Measures sample mass at different RH	Monitors particle morphology at different RH
(2) Sample status	Particles deposited on substrates	Particles deposited on substrates	Particles deposited on substrates	Particles deposited on substrates
(3) Size range	Not applicable	Not applicable	Not applicable	>1 μm
(4) Supersaturated samples	No	No	No	Yes
(5) Nonspherical particles	Yes	Yes	Yes	No
(6) Water adsorption	No	Yes	Yes	No
(7) Ambient application	Yes (off-line)	Yes (off-line)	Yes (off-line)	Yes (off-line)
	Electron microscopy	**AFM**	**X-ray microscopy**	**FTIR spectroscopy**
(1) Working principle	Monitors particle morphology at different RH	Monitors particle morphology at different RH	Monitors particle morphology at different RH	Monitors IR spectra of the sample at different RH
(2) Sample status	Particles deposited on substrates	Particles deposited on substrates	Particles deposited on substrates	Particles deposited on substrate
(3) Size range	>10 nm	>10 nm	>200 nm	Not applicable
(4) Supersaturated samples	Yes	Yes	Yes	Yes
(5) Nonspherical particles	No	No	No	Yes
(6) Water adsorption	No	No	No	Yes

Table 2.2 (continued)

(7) Ambient application	Yes (off-line)	Yes (off-line)	Yes (off-line)	Yes (off-line)
	Raman spectroscopy	**EDB**	**Optical levitation**	**Acoustic levitation**
(1) Working principle	Monitors Raman spectra of sample at different RH	Measures the mass of levitated particles as different RH	Measures diameters of levitated particles as different RH	Measures diameters of levitated particles as different RH
(2) Sample status	Particles deposited on substrates	Levitated particles	Levitated particles	Levitated particles
(3) Size range	Not applicable	A few to tens of μm	One to tens of μm	>20 μm
(4) Supersaturated samples	Yes	Yes	Yes	Yes
(5) Nonspherical particles	Yes	Yes	No	No
(6) Water adsorption	No	No	No	No
(7) Ambient application	Yes (off-line)	No	No	No
	H-TDMA	**Light extinction**	**Light scattering**	
(1) Working principle	Measures aerosol diameters at different RH	Measures aerosol light extinction at different RH	Measures aerosol light scattering properties at different RH	
(2) Sample status	Aerosol particles	Aerosol particles	Aerosol particles	
(3) Size range	<1 μm	A few nm to a few μm	A few nm to a few μm	
(4) Supersaturated samples	Yes	Yes	Yes	
(5) Nonspherical particles	No	No	No	
(6) Water adsorption	No	No	No	
(7) Ambient application	Yes (off-line)	Yes (off-line)	Yes (off-line)	

Note: Reproduced with permission from reference [5] with permission under the Creative Commons Attribution 4.0 CC-BY License, © Author(s) 2019. QCM, quartz crystal microbalance; AFM, atomic force microscopy; EDB, electrodynamic balance; TDMA, tandem differential mobility analyser.

A number of techniques were developed to investigate heterogeneous IN of atmospheric aerosol particles. In mixed-phase clouds that exist in $-38 < T < 0$ °C, immersion freezing is the most relevant mechanism in IN particles [28]. To explore the sensitivity and accuracy of these techniques, Hiranuma et al. [6] published an intercomparison study of the results obtained for the immersion freezing behavior of illite NX particles using different IN methods listed in Table 2.3. This reference material is illite-rich and mixed with a range of other minerals and was chosen as a surrogate for natural mineral dust. The bulk composition determined using X-ray diffraction analysis (XRD) shows that it is composed of 69% illite, 14% feldspar (orthoclase/sanidine), 10% kaolinite, 3% quartz, and 3% calcite/carbonate. The surface area using N_2 and H_2O vapor as adsorbing gases was measured to be 124 m^2 g^{-1} for both gases. Illite NX was distributed to researchers within the framework of the Ice Nuclei research UnIT (INUIT) where experiments were completed using 17 techniques designed for suspension or dry particle IN measurements (Table 2.3). The experiments were conducted over a wide range of particle concentrations, temperatures, cooling rates, and nucleation times. Immersion freezing activities were surface-area scaled, and their trends were correlated with sample preparation and particle characterization parameters. For comparison of results using continuous-flow diffusion chamber (CFDC), Garimella et al. [7] reported results on homogeneous and heterogeneous freezing experiments using a commercial spectrometer for ice nuclei (SPIN) that has a CFDC and an optical particle counter. SPIN measurements using illite NX covered the portion of the cirrus cloud regime and also captured the temperature dependence of ice activity. More details on specific examples relevant to atmospheric aerosols are presented in Chapter 3.

2.1.2 Mixing states and morphologies

The complex chemical composition of atmospheric aerosol particles influences their physical properties such as morphologies. Therefore, experimental measurements of the physical and chemical properties provide information on their distribution within the particle population of an aerosol, that is, physicochemical mixing state (see Chapter 1 for an expanded definition). Riemer et al. [8] and Laskin et al. [9] reviewed the literature on the techniques used to obtain information on aerosol's mixing state, and their advantages and limitations. Figure 2.2A shows a summary diagram of these techniques with their key features. The former review classified instruments suitable for these measurements to three groups: electron/X-ray microscopy, vibrational spectroscopy, and mass spectrometry. Imaging single particles using scanning electron microscopy (SEM) and transmission electron microscopy (TEM) techniques provides qualitative information on mixing state and morphology. When coupled with EDX and proper calibration with standard samples, quantitative information on elemental composition can be obtained. Interfacing the TEM setup with EELS provides information on chemical bonding. The use of synchrotron light in

Table 2.3: Summary of Ice Nuclei research UnIT (INUIT) measurement techniques and instruments.

ID	Instrument	Description	Portable?	Reference	Investigable T range	Ice detected T range for this study
		Suspension measurements				
1	BINARY*	Cold stage-supported droplet assay	No	[85]	−25 °C < T < ∼ 0 °C	−24 °C < T < −15 °C
2	CSU-IS	Immersion mode ice spectrometer	Yes	[86]	−30 °C < T < ∼ 0 °C	Poly: − 25 °C < T < −11 °C Mono: − 26 °C < T < −20 °C
3	Leeds-NIPI	Nucleation by immersed particle instrument	No	[87]	−36 °C < T < ∼ 0 °C	−21 °C < T < −11 °C
4	M-AL*	Acoustic droplet levitator	No	[88]	−30 °C < T < ∼ 0 °C	−25 °C < T < −15 °C
5	M-WT*	Vertical wind tunnel	No	[89, 90]	−30 °C < T < ∼ 0 °C	−21 °C < T < −19 °C
6	NC State-CS	Cold stage-supported droplet assay	No	[91]	−40 °C < T < ∼ 0 °C	−34 °C < T < −14 °C
7	CU-RMCS	Cold stage-supported droplet assay	No	[92]	−40 °C < T < −20 °C	−32 °C < T < −23 °C
12	FRIDGE*	Substrate-supported diffusion and condensation/immersion cell	Yes	[93]	−25 °C < T < −8 °C	[a]Default: − 25 °C < T < − 18 °C [b]Imm.: − 25 °C < T < −18 °C
		Dry particle measurements				
8	AIDA*	CECC	No	[94] [95, 96]	−100 °C < T < −5 °C	Poly: − 35 °C < T < −27 °C Mono: − 34 °C < T < −28 °C
9	CSU-CFDC	Cylindrical plate CFDC	Yes	[97]	−34 °C < T < −9 °C	−29 °C < T < −22 °C
10	EDB*	Electrodynamic balance levitator	No	[98]	−40 °C < T < −1 °C	[c]Imm.: − 31 °C < T < −28 °C [d]Contact: − 34 °C < T < −27 °C
11	FINCH*	Continuous-flow mixing chamber	Yes	[99]	−60 °C < T < −2 °C	−27 °C < T < −22 °C

(continued)

Table 2.3 (continued)

ID	Instrument	Description	Portable?	Reference	Investigable T range	Ice detected T range for this study
13	LACIS[*]	Laminar flow tube	No	[100, 101]	$-40\,°C < T < -5\,°C$	$-37\,°C < T < -31\,°C$
14	MRI-DCECC	Dynamic CECC	No	[102]	$-100\,°C < T < \sim 0\,°C$	Poly: $-26\,°C < T < -21\,°C$ Mono: $-29\,°C < T < -21\,°C$
15	PINC	Parallel plate CFDC	Yes	[103, 104]	$-40\,°C < T < -9\,°C$	$-35\,°C < T < -26\,°C$
16	PNNL-CIC	Parallel plates CFDC	Yes	[105]	$-55\,°C < T < -15\,°C$	$-35\,°C < T < -27\,°C$
17	IMCA-ZINC	Parallel plate CFDC	No	[106] [107, 108]	$-65\,°C < T < -5\,°C$	[e]Imm.: $-36\,°C < T < -31\,°C$ [f]Zinc: $-33\,°C < T < -32\,°C$

Notes: [*]Instruments of INUIT project partners, [a]default deposition nucleation, [b]immersion freezing with suspended particles, [c]immersion freezing, [d]contact freezing, [e]immersion freezing with IMCA, [f]Zinc alone. Acronyms: BINARY, Bielefeld Ice Nucleation ARraY; CSU-IS, Colorado State University Ice Spectrometer; Leeds-NIPI, Leeds Nucleation by Immersed Particles Instrument; M-AL, Mainz acoustic levitator; M-WT, Mainz vertical wind tunnel; NC State-CS, North Carolina State cold stage; CU-RMCS, University of Colorado Raman microscope cold stage; FRIDGE, FRankfurt Ice Deposition freezinG Experiment; AIDA, aerosol interaction and dynamics in the atmosphere; CSU-CFDC, Colorado State University continuous-flow diffusion chamber; EDB, electrodynamic balance; FINCH, Fast Ice Nucleus CHamber; LACIS, Leipzig Aerosol Cloud Interaction Simulator; MRI-DCECC, Meteorological Research Institute DCECC; PINC, portable ice nucleation chamber; PNNL-CIC, Pacific Northwest National Laboratory Compact Ice Chamber; IMCA-ZINC, Zurich Ice Nucleation Chamber with Immersion Mode Cooling chAmber; CECC, controlled expansion cloud-simulation chamber; CFDC, continuous-flow diffusion chamber. Note "poly" and "mono" denote polydisperse and quasi-monodisperse size-selected particle distributions, respectively. Reproduced and modified with permission from reference [6] under CC Attribution 3.0 License CC-BY, © Author(s) 2015.

(A)

Measurements Techniques for
Mixing States (Single particle)

1- Category:	Microscopy	Vibrational Spectroscopy	Mass Spectrometry
2- Information:	Spatial distribution (elements, species) Morphology	Composition Chemical bonding (functional groups)	Composition (Molecules and fragments)
3- Examples:	TEM (-EDX, -EELS) SEM (-EDX) STXM-NEXAFS AFM, AFM-IR XFM SIMS NanoSIMS Optical Micro-FTIR or Raman	ATR-FTIR EPMA/ATR-FTIR Raman SERS/TERS Fluorescence	PALMS ATOFMS RSMS SPLAT ALABAMA LAMPAS NAMS LMMS

(B)

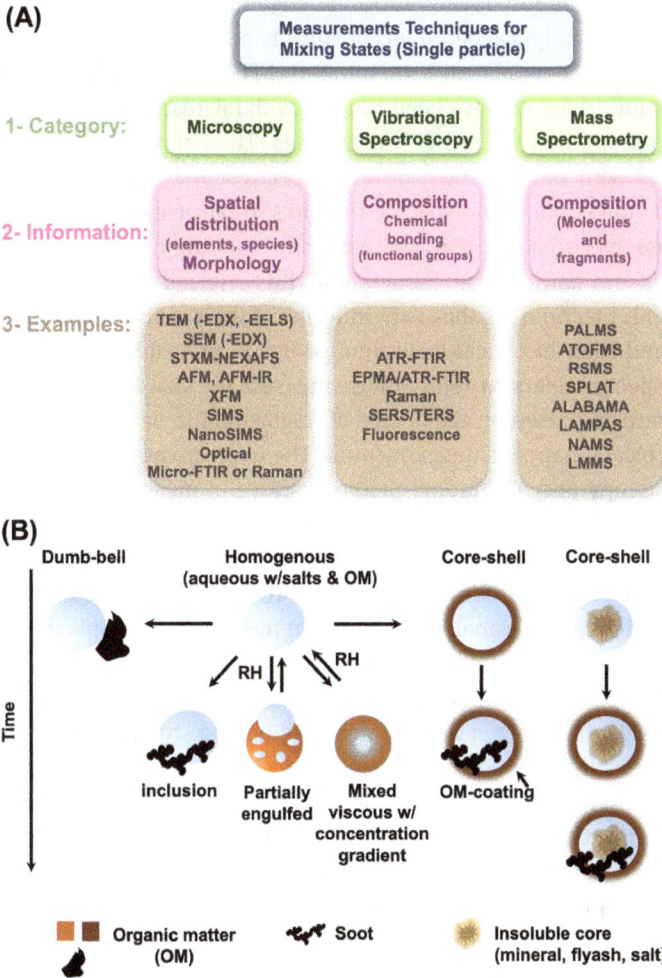

Dumb-bell   Homogenous (aqueous w/salts & OM)   Core-shell   Core-shell

Time

RH ↕   RH

inclusion   Partially engulfed   Mixed viscous w/ concentration gradient   OM-coating

Organic matter (OM)   Soot   Insoluble core (mineral, flyash, salt)

Figure 2.2: (A) Classification of single-particle techniques used to measure mixing states in a population of aerosol according to their category and the information they provide per Riemer et al. [8] and Laskin et al. [9]. Acronyms for the examples: TEM, transmission electron microscopy; EDX, energy-dispersive X-ray microanalysis; EELS, electron energy-loss spectroscopy; SEM, scanning electron microscopy; STXM, scanning transmission X-ray microscopy; NEXAFS, near-edge fine structure spectroscopy; AFM, atomic force microscopy; IR, infrared; TOF-SIMS, time-of-flight secondary ion mass spectrometry; EPMA, energy probe X-ray microanalysis; ATR-FTIR, attenuated total reflectance-Fourier-transform infrared spectroscopy; SERS, surface-enhanced Raman spectroscopy; TERS, tip-enhanced Raman spectroscopy; PALMS, particle analysis by laser mass spectrometry; ATOFMS, aerosol time-of-flight mass spectrometry; RSMS, rapid single-particle mass spectrometry; SPLAT, single-particle laser ablation time-of-flight mass spectrometer; ALABAMA, aircraft-based laser ablation aerosol mass spectrometer; LAMPAS, laser mass analysis of particles in the airborne state; NAMS, nanoaerosol mass spectrometer; and LMMS, laser microprobe mass spectrometry. (B) Schematic of major atmospheric aerosol mixing structures per You et al. [12], Li et al. [29], and Song et al. [30].

scanning transmission X-ray microscopy (STXM)/near-edge fine structure spectroscopy (NEXAFS) improves the analytical depth on chemical bonding. Vibrational spectroscopies such as Raman and Fourier-transform infrared (FTIR) provide information on functional group composition and have been coupled with microscopy techniques in Raman microspectroscopy and micro-FTIR techniques. The coupling of atomic force microscopy (AFM) with photothermal infrared spectroscopy in AFM-IR was demonstrated to detect functional group-specific IR peaks in particles smaller than 100 nm. Data from mass spectrometry increased our understanding of the chemical composition of aerosols and particle size. Riemer et al. [8] reviewed single-particle mass spectrometry (SPMS) techniques that vary in terms of their ionization source and mass analyzers, and their historical evolution for use in atmospheric aerosol characterization. The spectrometers currently in use measure the size and chemical species in individual particles based on the degree of fragmentation in the chemical analysis step, which typically happens under vacuum. The main advantage of SPMS compared to the microscopy and vibrational methods above in measuring mixing states is the much higher throughput that generates greater particle statistics needed for detailed understanding of the aerosol population [8]. In addition, for spatial distribution, time-of-flight secondary ion mass spectrometry (TOF-SIMS) and nano-SIMS were coupled with individual particle analysis to highlight their potential application. Table 1 in Laskin et al. [9] compares images from these mass spectrometric techniques to those from other particle imaging methods. Using some of these methods, You et al. [12], Li et al. [29], and Song et al. [30] identified mixing structures beyond homogeneous-like particles that include dumb-bell, core–shell, and organic matter (OM) coating as shown in Figure 2.2B. These mixing structures depend on particle size and are not static and tend to evolve over time during multiphase aging processes in the atmosphere including variation in RH, which can lead to liquid–liquid-phase separation and other phase transitions as discussed further [12, 30].

2.1.3 Viscosity

Viscosity is a physical property of fluids that characterizes their fluidity or stiffness [31, 32]. Values of the dynamic viscosity (η) varies by orders of magnitude as illustrated in Figure 2.3 for different materials. As a result, substances can be classified as liquids ($\eta < 10^2$ Pa s), semisolids ($10^2 < \eta < 10^{12}$ Pa s), and solids ($\eta > 10^{12}$ Pa s). For the latter, the "glass marble" was used as an example for amorphous solids with no long-range order. Hence, "glassy" state of matter exists when liquids undergo quench cooling and rapid drying, and when the vapor phase gets deposited at low temperatures [32]. As a result, the transition into a glass depends on the kinetics such as cooling or drying rates. In contrast with thermodynamic phase transitions that represent equilibrium conditions, the "glass transition temperature", T_g, represents a nonequilibrium transition. Koop et al. [32] provided detailed

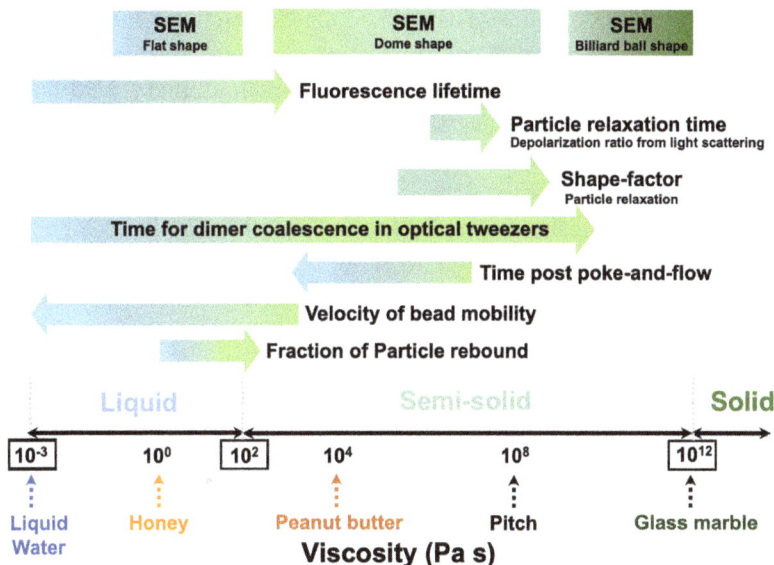

Figure 2.3: A diagram summarizing the trend in the measured quantities in techniques used for viscosity determination of lab-generated and field-collected aerosol particles as per reference [11].

analysis of how T_g can be used as a reference point for the estimation of viscosities in organic aerosol (OA) particles because viscosity is proportional to T_g. They also analyzed the dependency of T_g on parameters and processes relevant to OA particles that include molar mass, degree of oxidation, interactions in mixtures, adduct and oligomer formation, and water uptake as a function of RH. Examples are provided in Chapter 3.

The viscosity of liquids affects molecular transport processes, such as self-diffusion. The diffusion coefficient, D (in $cm^2\ s^{-1}$), of a molecule that makes up the bulk of the liquid is related to η through the Stokes–Einstein equation (2.1) [32]:

$$\eta = \frac{k_B T}{6\pi D r} \tag{2.1}$$

where k_B is the Boltzmann constant, T is temperature, and r is the apparent radius of the molecule which is assumed to be spherical. The inverse relationship between η and D suggests that as particles transition from liquid to semisolid and solid states which span 15 orders of magnitude increase in η, a reduction in the values of D will be observed in orders of magnitude as well. Measurements of viscosity for a given material allow for the estimation of D via eq. (2.1) [33–35]. For example, for a 10^{-8} cm molecular radius of an organic molecule at 298 K, values of D would be $>10^{-10}\ cm^2\ s^{-1}$ for liquid, 10^{-10}–$10^{-20}\ cm^2\ s^{-1}$ for semisolid, and $\leq 10^{-20}\ cm^2\ s^{-1}$ for solid physical states using the ranges of η values above.

While the viscosity of SOA particles has been assumed to be liquid in gas–particle partitioning models, experimental evidence using a variety of techniques has shown that they could be in semisolid or solid states under ambient conditions [15, 36, 37]. In their review of the phase of organic particles [11], Reid et al. provided a comprehensive synthesis of the literature of field observations and laboratory studies on the viscosity ranges of these particles since it is challenging to directly measure particle viscosity. Figure 2.3 shows a summary of the techniques used for correlating measured parameters with the range of particle viscosities. These techniques are briefly presented below with key references for further reading referring to the top of the figure. Specific results from each technique are provided in Chapter 3.

SEM imaging of single particles in the 0.2–3 µm size range was reported [38–40] to provide qualitative distinction over the viscosity range shown in Figure 2.3. Field and lab-generated particles were found to deform into different shapes related to their viscosity and surface tension upon water uptake: flat for liquid, dome for semisolid, and billiard ball shape for solid. For example, Figure 2.4 shows SEM images of SOA particles from the ozonolysis of α-pinene collected at low humidity (left panel, <5% RH), and after exposure to 80% RH followed by drying to <5% (right panel) [40]. The agglomeration of monomers is highlighted by red circles. The nonspherical agglomerates observed for the viscous SOA at low RH changed shape to spherical particles of lower viscosity following water uptake at high RH.

Fluorescence lifetime of molecular probes, referred to as "molecular rotors", depends on the viscosity of their environment. The technique based on this phenomenon is called fluorescence lifetime imaging, which has the resolution to study particles in 0.5–100 µm range. The measured fluorescence lifetime (τ_f) for some of these special classes of molecules increases linearly with increasing viscosity in the range 10^{-3}–10^{3} Pa·s according to the Förster–Hoffmann equation (2.2) [41]:

$$\log \tau_f = log\left(\frac{z}{k_r}\right) + \alpha \log \eta \tag{2.2}$$

where k_r is the radiative rate constant, and z and α are constants obtained from calibration experiments, where the molecular probe is added to solutions of known viscosity. For molecular rotors that do not follow eq. (2.2), empirical equations are derived, which best describe the dependency of τ_f on η [41].

In the light scattering experiments, which were performed in the European Organization for Nuclear Research (CERN) Cosmics Leaving OUtdoor Droplets (CLOUD), a new in situ depolarization instrument was developed (SIMONE-Junior) [42]. The depolarization signal from particles scattering light is sensitive to shape changes, from nonspherical for viscous particles to spherical in liquid and low viscosity particles. This signal is used to calculate a depolarization ratio, which is higher for nonspherical shapes and decreases to a constant, zero level, for optically spherical shape. Shape transitions occurred as a function of RH and temperature (−38 to +10 °C),

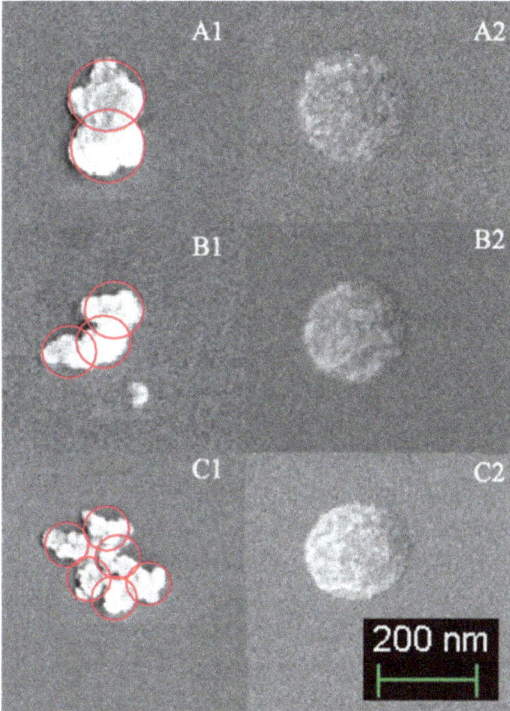

Figure 2.4: SEM images of SOA particles from the ozonolysis of α-pinene at low RH (<5%, left panel A1–C1) and the corresponding particles after increasing RH to 80% followed by drying to <5% (A2–C2). Red circles identify the monomer in the agglomerates. Reproduced with permission from [40] under CC Attribution 3.0 License CC-BY, © Author(s) 2015.

which allow for the measurements of viscosity/shape transition RH from the changes to the depolarization ratio. This depolarization-based RH was found to increase with decreasing temperature for SOA particles from the oxidation of α-pinene (~0.1–0.9 μm), which correlates with increasing particle viscosity in the semisolid range (~10^7 Pa·s). With this technique, the time for particle relaxation to a new shape can be used to roughly estimate viscosity based on the method described by Pajunoja et al. [39] for coalescing particles. Equation (2.3) relates the coalescence relaxation time (τ) to η [39]:

$$\tau = \frac{\eta\,d}{\sigma} \tag{2.3}$$

where d is the diameter of the primary particle and σ is the surface tension.

Moreover, for probing shape relaxation due to viscosity changes, Martin et al. [40] demonstrated the application of differential mobility analyzer–aerosol particle mass analyzer (DMA-APM) for probing shape changes of suspended organic particles with increasing RH, and using the results to estimate viscosity with the aid of numerical simulations. In the DMA-APM system, a condensation particle counter (CPC) was used to measure number-mass (m_p) distribution, and a scanning mobility particle sizer (SMPS) was used to measure the number-diameter (d_m) distribution of the particle population. This information was used to calculate a dimensionless dynamic shape factor, χ, according to the following equation [40]:

$$\chi = \frac{d_m}{(6m_p/\pi\rho)^{1/3}} \frac{C_C\left((6m_p/\pi\rho)^{1/3}\right)}{C_C(d_m)} \tag{2.4}$$

where ρ is the material density and C_C is the Cunningham slip correction factor. For spherical shapes, χ is near unity and is >1 for dimers, trimers, and higher order agglomerates. Therefore, χ values decrease with increasing RH as viscous particles change shape from aspherical to spherical. This shape relaxation process occurs within a specific time period that can be used to estimate viscosity using a commercial package for numerical solutions to fluid flows.

Holographic aerosol optical tweezers [43] were also used to monitor coalescence events of dimers to form spheres in the size range from 5 to 20 μm with increasing RH over time scales spanning 10^{-6} to 10^5 s depending on viscosity in the range from 10^{-3} to 10^9 Pa·s. This technique is based on the measurements of the scattered light intensity that probe the capillary-driven relaxation of the composite droplet as a function of time. The coalescence timescale, τ, is proportional to η via the following equation [43]:

$$\tau = \frac{2(2\ l^2 + 4\ l + 3)\ r\ \eta}{l(l+2)(2l+1)\ \sigma} \tag{2.5}$$

where l is the sum of the decaying normal modes of oscillation, r is the radius, and σ is the surface tension.

For estimating viscosity of microsized particles deposited on substrates by an impactor, the poke flow and the bead mobility techniques were used. In the poke flow experiments [44], semisolid particles in the 25–70 μm size range are poked with a needle attached to a micromanipulator to generate a nonequilibrium state. The time required for the material flow, $t_{(exp,flow)}$, is to restore the equilibrium driven by the surface energy of the system. Values of $t_{(exp,flow)}$ are determined experimentally and are inversely proportional to viscosity. Along with other physical properties of the particles such as particle shape, size, and temperature, values of $t_{(exp,flow)}$ are used to estimate viscosities through numerical simulations.

The bead mobility technique is based on measuring the velocity of 1 μm melamine beads incorporated in large particles (30–50 μm) deposited on a substrate as a result of flowing gas over the particle surface from the positions monitored with an optical microscope [45]. A calibration curve of bead velocity versus viscosities of standard systems is used to extract viscosity values for atmospherically relevant particles for up to 10^3 Pa·s.

Lastly, the use of the particle bouncing technique demonstrated that secondary particles can exist in semisolid and solid states under ambient conditions [46]. In this technique, two SMPS instruments and one electrical low-pressure impactor that is a twelve-state cascade impactor allowed for size-dependent measurements of electrical current. The "bounce factor" was defined as shown in the following equation [46]:

$$\text{Bounce factor} = \frac{\substack{\text{"excess" current resulting from the charges carried} \\ \text{by the bounced particles in the lowest impactor stage} \\ \text{that detects the smallest size particles}}}{\text{idealized signal inferred from SMPS data}} \tag{2.6}$$

This system was tested with samples with known viscosity such as dioctyl sebacate liquid droplets, crystalline ammonium sulfate particles, and amorphous solid polystyrene particles. Significant bounce was observed for the latter two samples compared to negligible results for the first system.

Bateman et al. [47] developed another apparatus to obtain particle rebound fraction, f. The apparatus consists of a three-arm impactor: a rebound arm containing an uncoated impaction plate, a capture arm containing a viscous-liquid-coated impaction plate, and a null arm with no impaction plate. A CPC was placed downstream of each arm to measure particle number concentrations. Computational fluid dynamics simulation and particle trajectories upon impaction allowed for quantifying the rebound fraction. This apparatus allows for measurements as a function of RH. Viscosity calibration (rebound fraction versus particle viscosity) using sucrose particles showed how the recorded data as a function of RH is associated with viscosity for two particle sizes, 190 and 240 nm (Figure 2.5). The transition from semisolid to liquid in the 25–80% RH is visible as a sharp decrease in f. For RH <25%, f values were found to be higher than unity due to fracturing of sucrose particles upon impaction. The f versus RH curves obtained using SOA from the oxidation of α-pinene, toluene, and isoprene suggested semisolid to liquid transitions over the 10^0–10^2 Pa·s range at characteristic RH values implying quantitatively different interactions of water with each type of SOA. For example, the RH for complete adhesion ($f = 0$) is larger than 65% for isoprene SOA, and 10:1 isoprene/α-pinene SOA compared to 95% for α-pinene SOA from dark ozonolysis and photo-oxidation using H_2O_2.

2.1.4 Optical properties

The quantities that quantify optical properties of atmospheric aerosol particles are the real (n, responsible for light scattering) and imaginary (k, responsible for light absorption) components of the material complex refractive index, $m = n - ik$. Both components contribute to the extinction coefficient of the material of interest [48]. These quantities vary with the wavelength, which for atmospheric aerosol particles,

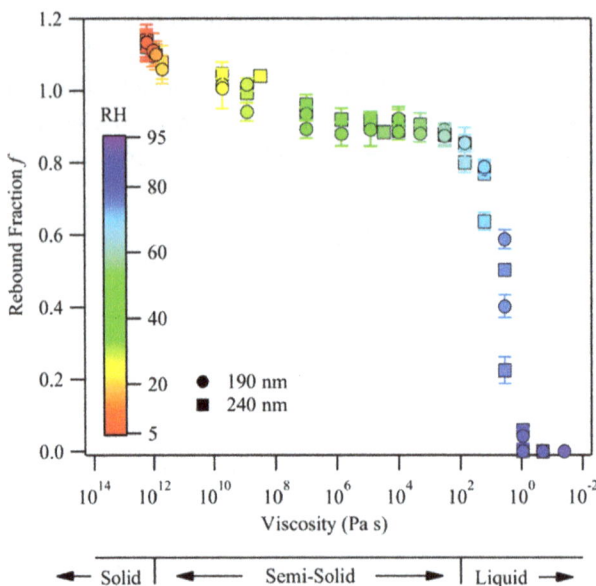

Figure 2.5: Viscosity calibration of rebound fraction, f, for sucrose particles with two particle mobility diameters as a function of RH. Reprinted (adapted) with permission from reference [109], © 2015 American Chemical Society.

ranges from the near-UV to the IR part of the electromagnetic spectrum. As outlined in the review article by Laskin et al. [10], the measurements of optical properties will vary depending on the sample being analyzed: bulk solid aerosol materials as thin films, bulk liquid aerosol extracts, air containing aerosols, and aerosol-loaded filters.

Commercial UV–vis [10] and FTIR [49] spectrophotometers enable the collection of extinction spectra in the lab for a wide range of materials. The technical challenge is with sample preparation, detector sensitivity, and reference materials that will make it possible to isolate scattering contributions to the measured extinction. According to the Beer–Lambert law, the attenuation of radiation, $I_0(\lambda)$, propagating through a homogeneous thin film with thickness, l, is related to the wavelength-dependent extinction coefficient, $\alpha_{bulk}^{ext}(\lambda)$, of the bulk material according to the following equation [10]:

$$\frac{I(\lambda)}{I_0(\lambda)} = exp\left[-\alpha_{bulk}^{ext}(\lambda)l_{film}\right] \tag{2.7}$$

Acquiring a sample spectrum relative to a nonabsorbing film with a similar real refractive index (n) would cancel out the scattering contribution. As a result, the measured extinction coefficient would be dominated by the absorption contribution, $\alpha_{bulk}^{ext}(\lambda) \approx \alpha_{bulk}^{abs}(\lambda)$. The value of l_{film} can be obtained from measurements or calculations, which introduces uncertainty. A technique like an optical surface profiler can yield thickness values in the micrometer range for a pressed film of a known mass and area. With this information, the material density, ρ, can be calculated by $\rho = \text{mass}_{\text{film}}/(\text{Area}_{\text{film}} \cdot l_{\text{film}})$, which is needed to calculate the mass absorption coefficient of the material, MAC(λ), according to the following equation [10]:

$$\text{MAC}(\lambda) = \frac{\alpha_{\text{bulk}}^{\text{ext}}(\lambda)}{\rho_{\text{bulk}}} = \frac{A_{10}^{\text{film}}(\lambda) \cdot \text{Area}_{\text{film}} \cdot \ln(10)}{\text{mass}_{\text{film}}} \qquad (2.8)$$

where $A_{10}^{\text{film}}(\lambda)$ is the measured base-10 absorbance. For organic carbon (OC) particles in an aerosol, $\text{MAC}_a(\lambda) = K\lambda^{-\text{AAE}}$ where K is a factor that includes aerosol mass concentration; λ is the wavelength (nm); and AAE is the absorption Angstrom exponent which is obtained from fitting to observed spectra and used as the basis for distinguishing different types of OC [10]. According to the modeling study by Lack and Cappa, AAE value for internally mixed OC/black carbon range is from 1 to 1.6 and that for BrC is greater than 1.6 [50].

If the sample is a solution extract of complex organic materials, the overall mass of the dissolved organics can be measured or estimated in a given solution volume. Hence, mass concentration, C_{mass}, is used in the calculation of MAC(λ) in units of grams of dissolved material per cubic centimeter of solution (g cm^{-3}) as follows [10]:

$$\text{MAC}(\lambda) = \frac{A_{10}^{\text{film}}(\lambda) \cdot \ln(10)}{l_{\text{solution}} \cdot C_{\text{mass}}} \qquad (2.9)$$

In the case of airborne particles, the extinction coefficient of an aerosol depends on mass concentration, size distribution, shapes, and mixing states [10]. In this case, the measured extinction coefficient of an air sample containing particles, $b_{\text{air}}(\lambda)$, is distinct from the α coefficient above because it is reported as a base-e extinction instead of base-10 and has contributions from the absorbance and scattering coefficients of the aerosolized particles and gases as follows [10]:

$$b_{\text{air}}^{\text{ext}}(\lambda) = b_a^{\text{abs}}(\lambda) + b_a^{\text{sc}}(\lambda) + b_g^{\text{abs}}(\lambda) + b_g^{\text{sc}}(\lambda) \qquad (2.10)$$

The $b^{\text{abs}}(\lambda)$ term is related to the mass-normalized absorption cross section (or efficiency) of aerosol particles, $\text{MAC}_a(\lambda)$ or MAE(λ), and the particle mass concentration in air, C_{PM}, as follows [10]:

$$b_a^{\text{abs}}(\lambda) = \text{MAC}_a(\lambda) \cdot C_{\text{PM}} \qquad (2.11)$$

Aerosol-loaded filters are often analyzed for their optical properties by measuring light transmission, I, through the sample to that of a clean filter, I_0. As summarized in Laskin et al. [10], examples of these instruments include particle soot absorption photometer, aethalometer, multiangle absorption photometer, and multiwavelength absorbance analyzer. The extinction coefficient, $b_{\text{filter}}^{\text{ext}}(\lambda)$, for particles collected on a filter after sampling a certain amount of air volume is given by [10]

$$b_{filter}^{ext}(\lambda) = \frac{Area_{filter}}{Volume_{air}} \ln\left(\frac{I_0}{I}\right) \tag{2.12}$$

Similar to the aforementioned extinction coefficients, the value of $b_{filter}^{ext}(\lambda)$ depends on the aerosol absorption and scattering coefficients, $b_a^{abs}(\lambda)$ and $b_a^{sc}(\lambda)$ [10]. The implicit assumption in these values is that aerosol morphology remain the same on the filter as in air [10]. For example, nonabsorbing aerosols such as α-pinene SOA appeared absorbing on filters [51], and that brown carbon aerosol (BrC) forms liquid-like elongated beads when coating quartz filter fibers [52].

Other techniques were also developed that directly measure absorption, scattering, and extinction coefficients of aerosols. These methods and their basic principle of operation were summarized in Laskin et al. [10]. Briefly, cavity ring-down (CRD) spectroscopy, which gives nearly identical results with cavity-enhanced spectroscopy, is based on measuring the radiation decay time at a given wavelength using a photomultiplier tube placed behind an optical cavity. This optical cavity is constructed from a sample container with two carefully aligned and high reflective mirrors at the two ends of the container. Equation (2.13) shows the relationship between the decay times for the aerosol sample (τ) and the buffer gas (τ_0) and the extinction coefficient of an aerosol, α_a^{ext}:

$$\alpha_a^{ext} = \frac{1}{c}\left(\frac{1}{\tau} - \frac{1}{\tau_0}\right) \tag{2.13}$$

where c is the speed of light.

The CRD setup is coupled with integrated sphere nephelometry (ISN) in the "albedometer" [53, 54]. The ISN measures simultaneously the scattering coefficient at the same wavelength as the CRD. This approach results in higher precision values of the single scattering albedo, SSA, of an aerosol calculated from $b_a^{abs}(\lambda)$ and $b_a^{sc}(\lambda)$ defined above according to the following equation [10]:

$$SSA = \frac{b_a^{sc}}{b_a^{abs} + b_a^{sc}} \tag{2.14}$$

In Chapter 3, specific examples are presented on the measured optical properties of aerosol particles of various chemical composition.

2.1.5 Acidity

Determining the acidity of atmospheric aerosol particles, cloud water, and fog droplets is a prerequisite to fully understand their chemical reactivity, which changes their physical properties and overall impacts on air quality, ecosystems, and the climate. In their review on the current status of knowledge on the acidity of atmospheric condensed phases, Pye et al. [13] and Freedman et al. [14] summarized the literature findings and made a distinction between "aerosol pH" and "cloud pH" when discussing observations, measurements, and model estimations. Observations show that acidic fine particles are ubiquitous with pH as low as −1 due to the dominance of ammonium, sulfate, nitrate, and organics, in contrast with coarse-mode aerosols originating from sea spray and dust, which contain nonvolatile cations. Reductions in SO_2 and NO_x emissions over the past decades in the United States increased cloud/fog pH. Direct measurements of aerosol pH are challenging because of the liquid water content that is extremely high in ionic strength, low amounts of mass given the particles' submicrometer size, and chemical heterogeneity as a result of the complex mixing state of the particles. At the present time, the aerosol pH is based on thermodynamic model calculations, with best estimates obtained when gas–particle partitioning observations are taken into account. Although methods for determining the pH of bulk and individual aerosol samples continue to be developed, the following studies highlight the recent work in this area [13]:

2.1.5.1 Indirect proxy methods and thermodynamic equilibrium models

Hennigan et al. [55] provided a critical evaluation of proxy methods used to estimate aerosol acidity. According to Pye et al. [13], there are two major proxy methods based on the principle of solution electroneutrality: the cation/anion equivalent ratio and the charge/ion balance. Both methods require measuring the concentration of ions in solution using ion chromatography (IC) and aerodyne aerosol mass spectrometry (AMS). For the first method, one of the assumptions is that the charge deficit between the water-soluble species is accounted for by H^+. The ratio is calculated based on the concentration of the chemical species per unit volume of air (i.e., mol m^{-3}) as follows [13]:

$$\text{Cation/Anion} = \frac{\left[NH_4^+\right] + \left[Na^+\right] + \left[K^+\right] + 2\left[Ca^{2+}\right] + 2\left[Mg^{2+}\right]}{2\left[TSO_4\right] + \left[NO_3^-\right] + \left[Cl^-\right]} \qquad (2.15)$$

where TSO_4 refers to total sulfate, HSO_4^-, and SO_4^{2-}. Organic acids and the carbonate system influence the ion balance when the aerosols are near neutral.

In the second method, the total molar amount of H^+ per unit volume of air containing aerosol particles and/or cloud droplets needed for charge balance (cb) is related to pH and is expressed as follows [13]:

$$H_{air,cb}^+ = 2\left[TSO_4\right] + \left[NO_3^-\right] + \left[Cl^-\right] - \left(\left[NH_4^+\right] + \left[Na^+\right] + \left[K^+\right] + 2\left[Ca^{2+}\right] + 2\left[Mg^{2+}\right]\right) \quad (2.16)$$

Because the value of $H_{air,cb}^+$ is mass dependent, it has large uncertainties arising from the measurements of the aerosol-phase composition.

The critical analysis of proxy methods carried out by Pye et al. [13] led to the strong recommendation that proxies are to be avoided and encouraged the use of validated aerosol thermodynamic equilibrium models as one of the tools in study-ing particle acidity. Table 2.4 shows a summary of the features of each model and their advantages and disadvantages.

2.1.5.2 Direct bulk average pH measurements

For aerosol particles collected on filters, two methods were tested that utilize pH-sensitive indicators and pH paper. The first one involves spiking the filters with a pH-sensitive dye prior to aerosol collection. This method requires precise control of RH in the collection chamber [56]. UV–vis spectroscopy is used to analyze the indi-cator and quantify the concentration of the protonated versus the deprotonated spe-cies. The second method is referred to as the calorimetric method, which utilizes pH paper with a relatively high resolution (0.1 pH units) placed on multistage impac-tors [57]. Field aerosol particles or those generated in the lab are directly collected on the pH paper; hence, particle acidity is obtained as a function of size. The acidity of cloud and fog droplets was determined using the pH paper strips on impactor stages and correlated with the sulfur mass from a microchemical reaction of the droplets with calcite ($CaCO_3$) fragments that forms gypsum ($CaSO_4$) [58]. The calcite fragments were placed on the impactor stages as well. Micromatic crystals of gyp-sum were identified by SEM/EDS to determine the sulfur mass. The unreliability of these measurements originates from the high ionic strength that may affect the dye activity and the insufficient aerosol water that would substantially wet the pH paper for accurate color change.

2.1.5.3 Direct single-particle pH measurements

Recent advances in the direct measurements of pH in micrometer aqueous droplets have relied on changes in the spectral ratio of protonated versus deprotonated spe-cies of molecular pH indicators and acid–conjugate base pairs due to changes in pH. Table 2.5 lists the basis of operation for the most recent tools and methods used in lab studies. The advantages and disadvantages of these approaches have been summarized in table 1 in reference [59]. The suitability of these methods for field studies was not demonstrated yet. Nevertheless, the proof-of-concept results of each method opens the door for further analytical development and new lab experi-ments on microdroplets under controlled environments that could be designed to be atmospherically relevant.

The above discussion on pH estimation and measurements focused on aerosol particles, deliquesced particles, and microdroplets. As for pH in cloud and fog water, it is generally acidic to a lesser degree than particles, and its pH is sensitive to anthro-pogenic levels of ammonia, sulfur, and nitrogen oxides [13]. Because the sample vol-umes for fog/cloud and precipitation water that can be collected directly are orders of magnitude larger than for aerosol particles, alongside with dilute ion concentrations, conventional and well-established instrumentation is used for direct cloud water pH

Table 2.4: Common box thermodynamic equilibrium models used to calculate acidity.

Model	Input	Acidity output	Advantages	Disadvantages
E-AIM	Gas + particle or equilibrium particle composition (H^+, NH_4^+, Na^+, SO_4^{2-}, HSO_4^-, NO_3^-, Cl^-, Br^- organic acids, and amines) in moles overall electroneutral conditions (see eq. (2.19) for Z) ; RH, and T.	pH at equilibrium.	pH via recommended eq. (2.1). Considered the most accurate inorganic thermodynamic model. Ionizing organic species (e.g., organic acids and amines) included.	Computationally intensive. T and RH restricted for some compositions to preserve accuracy.
AIOMFAC-GLE	Gas + particle or equilibrium particle composition (H^+, Na^+, K^+, NH_4^+, Mg^{2+}, Ca^{2+}, Cl^-, Br^-, NO_3^-, HSO_4^-, SO_4^{2-} and organic species and/or organic functional groups) in mol m^{-3} air for electroneutral conditions ; RH and T	pH at equilibrium.	pH via recommended eq. (2.1). Accounts for organic–inorganic interactions and liquid–liquid equilibrium in a consistent framework. Code publicly distributed through repository.	Limited support for solid–liquid equilibria of diverse inorganic salts (presently). Optimized for temperatures near 298 K, with limited accuracy for much colder atmospheric temperatures. Organic species do not ionize.
MOSAIC	Distinct gas and particle composition (H^+, NH_4^+, Na^+, Ca^{2+}, SO_4^{2-}, HSO_4^-, $CH_3SO_3^-$, NO_3^-, Cl^-, and CO_3^{2-}) in mol m^{-3} air; RH and T. Automatic adjustments applied to nonelectroneutral input particle-phase composition.	pH$_F$ by default (pH ± with modification) for each particle size bin (or mode) at each tie step while dynamically solving gas–particle mass transfer.	Provides size-resolved pH$_F$ and pH ± to account for compositional heterogeneity across particles of different sizes and origins. Does not require equilibrium assumption.	Gas–particle and solid–liquid equilibrium constants depend on temperature but activity coefficients are limited to 298.15 K.

(continued)

Table 2.4 (continued)

Model	Input	Acidity output	Advantages	Disadvantages
ISORROPIA II	Gas + particle or particle composition (TSO_4, TCl, TNO_3, TNH_4, Na, K, Ca, and Mg) in mol m^{-3} or µg m^{-3} air; RH and T. Automatic adjustments applied to nonelectroneutral input particle-phase composition.	pH_F by default ($pH_\pm$ with modification) at equilibrium.	Computationally efficient. Code has widespread public distribution and incorporation in CTMs.	Approximation employed (e.g., some activity coefficients treated as 1, minor species do not perturb equilibrium, higher default numerical tolerances). Segmented solution approach leads to discontinuous solution surface.

Notes: Equation (2.1) refers to $pH = -\log_{10}(a_{H^+}) = -\log_{10}(m_{H^+}\gamma_{H^+})$, which is the recommended IUPAC definition of pH. $pH_F = -\log_{10}(m_{H^+})$, which is the free-H$^+$ approximation of pH. $pH_\pm(H, X) = -\log_{10}(m_{H^+}\gamma_{\pm,HX})$. "T" in "$TSO_4$, TCl, TNO_3, etc., refers to total. Equation (2.19) refers to $Z = [TNH_4] + [Na^+] + [K^+] + 2[Mg^{2+}] + 2[Ca^{2+}] - [TCl] - [TNO_3] - 2[TSO_4]$.

Table 2.5: Summary of emerging approaches to probe single-particle acidity per Pye et al. [13] and later studies not included in their review.

Name	Methodology basis	Reference
Protonated/deprotonated fluorescent indicator method	Using confocal and optical microscopy, droplets in the size 10–30 μm containing Oregon Green 488 (pK_a ~ 4.6) to probe pH before and after liquid–liquid-phase separation in model SOA systems contain ammonium sulfate and polyethylene glycol. The ratio of the fluorescence intensity of deprotonated to protonated dye is calibrated to obtain pH values at individual points throughout the aerosol, while a single phase at high RH, and phase separated in the organic-rich phase.	[110]
Acid–conjugate base method	Using Raman microspectroscopy, this method calibrates the peak area for the acid and conjugate base to molar concentrations. The activity of the H$^+$ ion is determined from the acid dissociation constant, K_a, and activity coefficient calculations.	[111–113]
	Using surface-enhanced Raman spectroscopy (SERS), nanometer-sized pH probes (4-mercaptobenzoic acid–functionalized gold nanoparticle) in suspended microdroplets (~20 μm dia.) were used to correlate changes in the signal intensity of the carboxylate groups in the nanoprobes to pH in the core and the cross section of the droplet for comparison with bulk solution.	[114, 115]
	Using aerosol optical tweezer (AOT) coupled with Raman spectroscopy (for composition) and cavity-enhanced Raman scattering (CERS) (for size and refractive index), the coalescence of a trapped aqueous aerosol (~8 μm dia.) with smaller aerosols containing a strong acid was monitored and used to determine the pH before and after these events with the specific ion interaction theory.	[59]
	Using an ultrasonic levitator to acoustically levitate 1.2 μL droplets containing phenol red as the acid–base indicator coupled with visible spectroscopic measurements. Droplet images were collected using a CCD camera. The ratio of the peak absorbance of the basic species of phenol red (pK_a = 7.9) to that of the acidic species varies linearly with solution pH over the pH = 7–9 range.	[116]

Note: See table 1 in reference [59] for advantages and disadvantages of these methods.

measurements. Electronic pH meters with combination of glass electrodes have been used for decades. The electrodes can work with milliliter to microliter sample volumes. Buck et al. [60] emphasized key operational aspects on calibrating the pH electrodes using appropriate standard buffers with certain specifications for the relevant pH ranges.

2.2 Measuring chemical composition

Table 2.1 lists the instrumentation used for off-line and real-time measurements of aerosol chemical composition. Advances in analytical tools used to measure the chemical composition of ensemble average and single particles (suspended or deposited on substrates) in lab studies and field campaigns were reviewed earlier [4, 15–17]. In general, chemical composition investigations aim to obtain quantitative and qualitative information of the elemental composition and oxidation states, mineralogy, soluble inorganic and organic species, and organic functional groups. The majority of these techniques provide bulk information and some have surface sensitivity and specificity (discussed below). Figure 2.2A shows the microscopy and vibrational spectroscopy techniques used to study the mixing states of aerosol particles, which also provide information on chemical composition [4, 9]. These techniques have surface sensitivity to the chemical composition of single particles. Selected examples of heterogeneous reactions studied using these techniques will be presented in Chapter 3.

2.2.1 Organic aerosols

Advances in the analysis of OAs from primary and secondary sources have been the subject of a number of reviews [15–17]. Hallquist et al. [15] classified field-deployable techniques according to three characteristics: completeness, chemical resolution, and time/size resolution as graphically shown in Figure 2.6. The "perfect instrument", which currently does not exist, is the one that provides complete organic mass analysis, molecular ID of organic species, and high size (nm range)/time (<1 h) resolution. The majority of the techniques shown in Figure 2.6 are "off-line methods" with the exception of two "online methods": AMS and particle-into-liquid sample/water-soluble organic carbon (PILS-WSOC) using IC or total organic carbon (TOC) analyzer. The online techniques are capable of real-time measurements with resolution in minutes.

Following the review by Hallquist et al. [15], Johnston and Kerecman [16] reviewed AMS and other mass spectrometry techniques used off-line and online for molecular characterization of atmospheric OA. As highlighted by Zhang et al. [61],

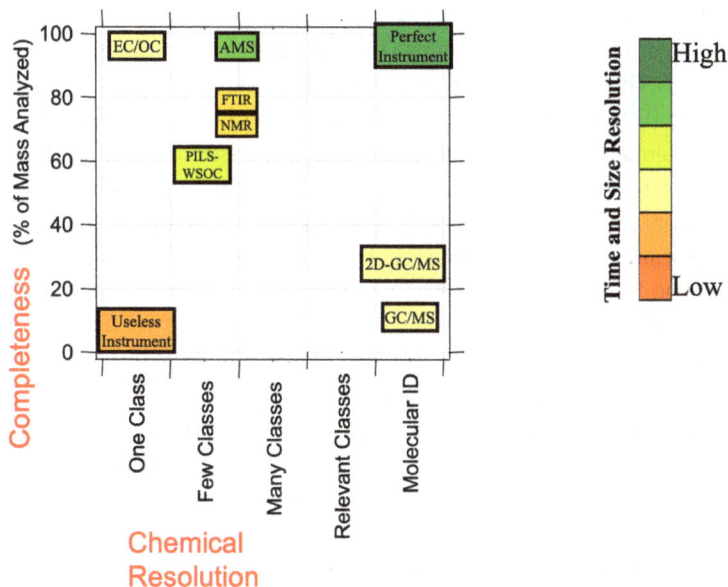

Figure 2.6: Three-dimensional representation of some of the techniques used for the analysis of the organic content in atmospheric aerosol particles to highlight their complementary nature. Abbreviations: EC/OC, elemental carbon/organic carbon; AMS, aerosol mass spectrometry; FTIR, Fourier-transform infrared spectroscopy; NMR, nucleic magnetic resonance; PILS-WSOC, particle-into-liquid sampler-water-soluble organic carbon; 2D-GC/MS and GC/MS, 2-D-gas chromatography/mass spectrometry. Reprinted from reference [15] with permission under the Creative Commons Attribution 3.0 License, © Author(s) 2009.

applying multivariate factor analysis techniques to AMS data provides quantitative and simplified description of individual organic species that contribute to the OA mass. The setup prior to introducing the samples to the mass spectrometer would require a thermodenuder to characterize organics with various degrees of volatility [16]. A setup designed to characterize chromophores would include high-performance liquid chromatography and a photodiode array detector as described by Laskin and co-workers [62, 63].

Timonen et al. [64] reviewed the application of online PILS techniques in the analysis of water-soluble compounds in field-collected aerosol from different field campaigns around the world. The shortest time resolution was 3 s and the longest was 2 h. For their study, they designed a new system, PILS-TOC-IC, where they coupled a commercial PILS with a TOC analyzer and two ion chromatographs for simultaneous high-time resolution measurements of soluble ions NO_3^-, SO_4^{2-}, Cl^-, NH_4^+, Na^+, K^+, Ca^{2+}, Mg^{2+}, and oxalate and water soluble organic carbon (WSOC). The TOC and IC time resolution of the measurements was 6 and 15 min, respectively. A high correlation level with $R > 0.9$ was found when comparing results of the

PILS-TOC-IC with those using high-resolution time-of-flight AMS (HR-TOF-AMS). Timonen et al. [64] also described a procedure for quantifying water insoluble organic carbon (WISOC). This method relies on subtracting the concentration of OC and elemental carbon (EC) from the measured WSOC. To obtain the concentration of OC and EC, a commercial thermal-optical carbon analyzer described in a previous study [65] was coupled to a cyclone with a cutoff at 1 μm that removed coarse particles and a carbon denuder to remove organic gaseous compounds from the sample air. OC was determined in the thermal method using helium gas with two temperature steps: 600 °C (for 80 s) and 840 °C (for 90 s). EC was determined in the thermal method using helium–oxygen mixture with three temperature steps: 550 °C (for 30 s), 650 °C (for 45 s), and 850 °C (for 90 s). Charring of OC was corrected optically using the data obtained from the tuned diode laser [64].

Examples of "off-line" techniques as per Figure 2.6 are FTIR and NMR. The application of infrared spectroscopy to the detection and quantification of functional groups in organic particles from smog chamber and field studies has been reviewed by Reggente et al. [66] and Gao et al. [67]. These reviews highlighted the work of researchers who used transmission FTIR and attenuated total reflectance ATR-FTIR for quantitative information. For example, Maria et al. [68] used a solubility behavior-based method [69] to obtain absorptivity and detection limits in μg cm^{-2} for inorganic ions and organic functional groups that include aliphatic, alkene, aromatic, and carbonyl carbon, in addition to alcohols, organosulfates, and amines. This type of analysis converted peak absorbances to moles of functional groups. Then, for field-collected samples, OC and organic matter (OM) were estimated using the following equations:

$$
\begin{aligned}
OC = \; & \left(0.5 \times \left[\text{moles alkane C} - \text{H}\right] \times \left[12\,\text{g mol}^{-1}\right]\right) \\
& + \left(\left[\text{moles alkene C} - \text{H}\right] \times \left[12\,\text{g mol}^{-1}\right]\right) \\
& + \left(\left[\text{moles aromatic C} - \text{H}\right] \times \left[12\,\text{g mol}^{-1}\right]\right) \\
& + \left(\left[\text{moles C} = \text{O}\right] \times \left[12\,\text{g mol}^{-1}\right]\right) \\
& + \left(\left[\text{moles C} - \text{OH}\right] \times \left[12\,\text{g mol}^{-1}\right]\right) \\
& + \left(\left[\text{moles C} - \text{NH}_2\right] \times \left[12\,\text{g mol}^{-1}\right]\right) \\
& + \left(\left[\text{moles C} - \text{O} - \text{S}\right] \times \left[12\,\text{g mol}^{-1}\right]\right)
\end{aligned}
\tag{2.17}
$$

$$OM = \left(0.5 \times [\text{moles alkane C} - \text{H}] \times [14 \text{ g mol}^{-1}]\right)$$
$$+ \left([\text{moles alkene C} - \text{H}] \times [13 \text{ g mol}^{-1}]\right)$$
$$+ \left([\text{moles aromatic C} - \text{H}] \times [13 \text{ g mol}^{-1}]\right)$$
$$+ \left([\text{moles C} = \text{O}] \times [28 \text{ g mol}^{-1}]\right) \qquad (2.18)$$
$$+ \left([\text{moles C} - \text{OH}] \times [29 \text{ g mol}^{-1}]\right)$$
$$+ \left([\text{moles C} - \text{NH}_2] \times [28 \text{ g mol}^{-1}]\right)$$
$$+ \left([\text{moles C} - \text{O} - \text{S}] \times [54 \text{ g mol}^{-1}]\right)$$

Dillner and Coury [70, 71] calibrated a method for using ATR-FTIR spectroscopy with multivariate calibration to quantify organic functional groups and inorganic ions (ammonium sulfate and ammonium nitrate) in size-segregated ambient samples for particles smaller than 1 µm. This calibration method [71] quantified 14 functional groups that include methyl (CH_3), methylene (CH_2), aliphatic (CH), alkene (C = C), aromatic (CH), aldehydes/ketones, carboxylic acids, esters/lactones, acid anhydrides, carbohydrate ether, carbohydrate hydroxyl, organic nitro group, amino acids, and amines. Standard compounds containing the functional groups of interest were used for the calibration with the expected concentration range. Then, calibration standards of compounds containing these classes of functional groups were created to mimic ambient spectra by varying the relative amounts of each class in each standard. The spectrum of each calibration standard and the known moles of each functional group in each calibration standard were used as the training set to develop a partial least squares multivariate calibration for the functional groups [71].

NMR spectroscopy is another off-line technique that has been used for the identification and quantification of functional groups in OA as reviewed by Duarte and Duarte [17, 72]. Figure 2.7 shows a diagram for the capabilities of different NMR techniques in deriving spectral and structural information. Solid-state and one- and two-dimensional (1-D and 2-D) solution-state NMR spectroscopies were demonstrated as effective tools for structural elucidation of OA from primary sources in field studies and secondary processes in lab investigations. The review highlighted the studies that utilized 1-D solid-state ^{13}C cross-polarization magic angle spinning NMR (CP-MAS NMR) for analyzing aerosol WSOM to obtain structural and semiquantitative information. These studies were motivated by the extensive literature on analyzing natural OM using ^{13}C CP-MAS NMR as a nondestructive technique with parameters that could be optimized for the maximum accuracy for quantitative analysis. These reviews [17, 72] also highlight the potential for using other nuclei besides carbon such as ^{15}N given the increasing importance of organonitrogen compounds as a class of brown carbon in atmospheric OA [73]. For solution-state NMR, Duarte and Duarte [17, 72] highlighted studies that analyzed WSOM using 1-D ^{1}H NMR spectroscopy

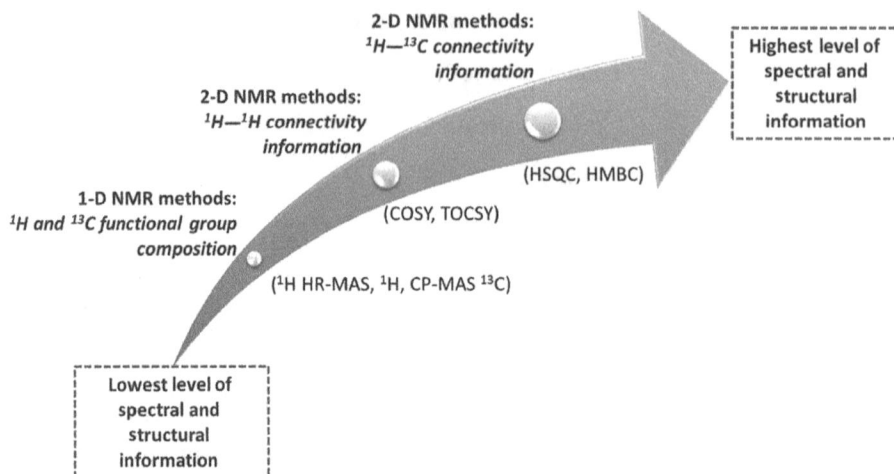

Figure 2.7: Structural information obtained from different NMR methods used in characterizing WSOM. Abbreviations: HR-MAS, high-resolution-magic angle spinning; COSY, correlation spectroscopy; TOCSY, total correlation spectroscopy; HSQC, heteronuclear single-quantum correlation; HMBC, heteronuclear multiple bond correlation. Reprinted from reference [17] with permission from © Elsevier, 2020.

using deuterated solvents, D_2O (with water suppression for HDO signal), dimethyl sulfoxide-$d6$ (DMSO-$d6$, δ_H 2.5 ppm), or methanol-$d4$ (MeOH-$d4$, δ_H 3.31 ppm). Quantitative analysis by the integration of proton peaks allows for making correlation graphs like the one shown in Figure 2.8. This figure shows a schematic representation for the distribution of functional groups with various degrees of saturation and oxygenation in WSOM from different sources. The boundaries shown in Figure 2.8 reveal interesting information such as the higher degree of unsaturation for SOA compared to fresh and aged BBOA and that bioaerosols like pollen have structures that separate them from combustion-based OA.

2.2.2 Inorganic aerosols

The inorganic fraction (elements, cations, anions, and trace metals) in collected atmospheric aerosol particles (dust and deliquesced aerosol) can be analyzed via online or off-line techniques. In section b of Table 2.1, the real-time measurements of the inorganic content are described in b2 and b3 for the sulfate, nitrate, and other inorganic species, respectively. Examples of the latest development in mass spectrometry techniques for chemical composition are listed in Figure 2.2A as per Riemer et al. [8] and Laskin et al. [9]. The reader is referred to these review articles for details on each of these mass spectrometry techniques. Duncan et al. [74] reviewed quantitative capabilities of

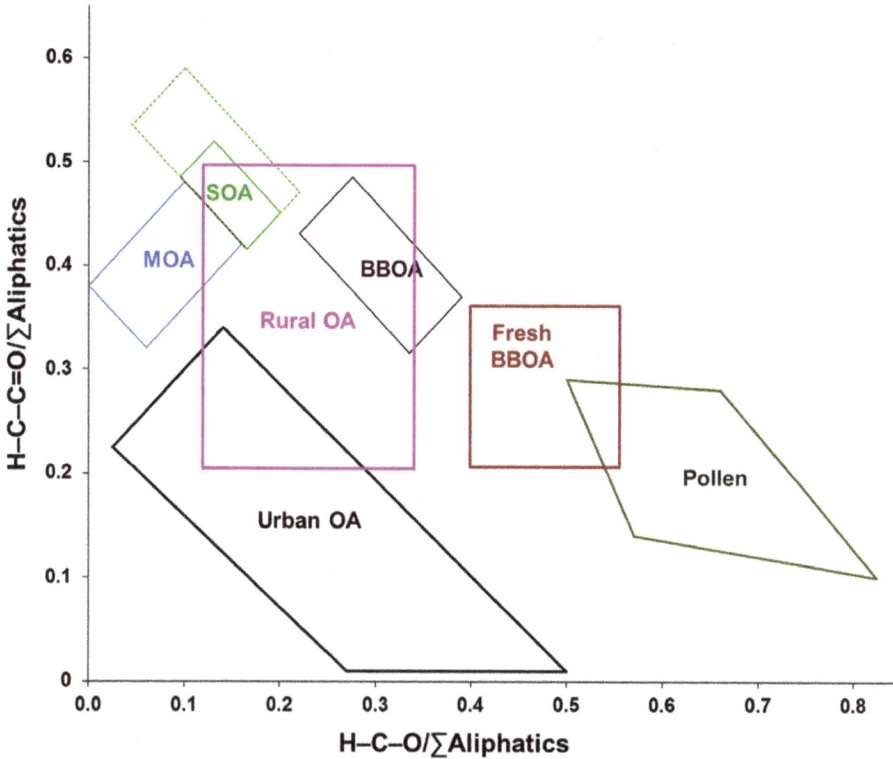

Figure 2.8: Functional group distribution in atmospheric WSOM from different sources from the quantitative analysis of ^{1}H-NMR spectroscopy as originally presented by Decesari et al. [117]. The axes represent the carbon-weighted ratios of H-C-C = O (to account for carbonylic/carboxylic groups) and O-C-H (to account for hydroxyl groups) to total aliphatic carbon fraction in the samples, which includes oxygenated function groups (saturated (H-C-O) and unsaturated (H-C-C = O)), benzylic groups (H-C-Ar), and the unfunctionalized alkyls (H-C). Abbreviations: MOA, marine organic aerosol; SOA, secondary organic aerosol; and BBOA, biomass burning organic aerosol. Reprinted from reference [17] with permission from © Elsevier, 2020.

mass spectrometry techniques in general and the important considerations for inter-preting measurements using these methods.

The off-line analysis depends on the solubility of the aerosol components in liquid water or other solvents [20]. The three major steps for samples considered for off-line analysis are collection, extraction, and analysis. Atmospheric aerosol particles can either be collected on filters, into a liquid sampler using a PILS instrument described above, or substrates for spatial elemental analysis. For size fractionation of collected particles, a micro-orifice uniform deposit impactor (MOUDI) is used, which contains multiple stages for a given particle size range [75]. The MOUDI is suitable for different kinds of filters or sample grids used in imaging for extraction and direct analysis of particles, respectively. The size fractionation of ultrafine particles (10–180 nm) is made possible using nanoMOUDI for chemical analysis [76].

The U.S. Environmental Protection Agency recommends certain procedures to be followed in their method IO-3.1 for the selection, preparation, and extraction of filter material to ensure consistency and reproducibility [77]. Extracts from control or blank filters are usually obtained for comparison. It is important to keep track of the pore size of the filters if extracts undergo filtration prior to analysis. This consideration is particularly important for solutes susceptible to colloidal formation and precipitation due to changes in pH such as trace metals. For example, Meskhidze et al. [78] used the term "soluble iron (SFe)" in deliquesced aerosol solution, cloud/rainwater, and seawater to refer to iron species that pass through a 0.02 µm filter. Once the extraction step is complete, whether from filters or using PILS, the extracts are then analyzed using standard analytical techniques such as IC, atomic emission spectroscopy, and inductively coupled plasma-mass spectrometry (ICP-MS) [20]. Method development and optimization for each of these techniques are required using standard solutions with concentrations in the range of unknowns, from ppt to lower ppm levels. For the analysis of trace metals, the protocols developed for GEOTRACES cruises is a very useful starting point since they contain details on best practices to avoid contamination during collection, storage, and handling of liquid samples [79].

For spatial elemental analysis [4, 9], electron and X-ray-based techniques are often used under various degrees of vacuum. Samples can be collected directly on substrates suitable for the imaging technique of interest such as those listed in Figure 2.2A. Quantitative elemental mapping using EDS is usually coupled with TEM and SEM. When properly calibrated, the EDS signal is converted to atomic% or weight%. Using synchrotron light sources, STXM/NEXAFS can also provide oxidation state and chemical bonding information. Analysis of metals and high atomic number elements in coarse particles is possible using X-ray fluorescence-based techniques that also provide spectral information (Table 2.6).

While spatially resolved data are useful for particle size and shape under "dry" conditions (with beam damage as the main drawback), other techniques used in the analysis of atmospheric dust particles are listed in Table 2.6 and include XRD for mineralogy and degree of crystallinity. The mineral composition in dust is often compared to that of source soils to assess fractionation that occurs during dust formation and transport [80, 81]. Dust particles are also often analyzed for elemental composition using ICP-MS for trace metal quantification.

The application of surface-sensitive techniques to probe the heterogeneous chemistry of field-collected and lab proxies of mineral dust, salts, and sea spray improved our understanding of changes to the surface chemical composition (elements and functional groups) and oxidation states under various reaction conditions [4, 19, 24, 80, 82–84]. Examples of these techniques are ambient pressure XPS, Auger electron spectroscopy, diffuse-reflectance FTIR spectroscopy, ATR-FTIR spectroscopy, and SFG. The reader is referred to the above reviews for details on their basis of operation.

Detailed examples of reactions studied in situ using these techniques are presented in the next chapters.

Table 2.6: Bulk techniques used for analyzing the chemical composition of atmospheric dust particles.

Technique name	Measured quantity	Unit	Ref.
X-ray fluorescence (XRF) – Micro-XRF (μ-XRF) – Total reflection XRF (TXRF) – Glazing angle synchrotron XRF (SR-XRF)	Elemental chemical composition and elemental mapping, semiquantitative	ppb–ppm ($\mu g\ kg^{-1}$–$mg\ kg^{-1}$), which can then be converted to mass per volume of air sampled	[118]
X-ray diffraction (XRD)	Mineralogy: identify crystalline phases of metal oxides, hydroxides, and carbonates. If diffraction intensity is calibrated with standards, weight fraction of each mineral can be calculated	wt%	[119]
Inductively coupled plasma-mass spectrometry (ICP-MS)	Elemental chemical composition: A known amount of the solid material is digested in strong acid (nitric acid–hydrochloric acid mixture) followed by analysis in the liquid phase	ppt–ppm ($ng\ L^{-1}$–$mg\ L^{-1}$), which can then be converted to mass per volume of air sampled	[120, 121]

References

[1]   McMurry PH. A review of atmospheric aerosol measurements. Atmos Environ. 2000;34:1959–1999.

[2]   Kulkarni P, Baron PA, Willeke K, editors. Aerosol Measurement: Principles, Techniques and Applications. 3rd ed, New Jersey: Wiley; 2011.

[3]   Signorell R, Reid JP, editors. Fundamentals and Applications in Aerosol Spectroscopy. 1st ed, Boca Raton: CRC Press; 2011.

[4]   Ault AP, Axson JL. Atmospheric aerosol chemistry: Spectroscopic and microscopic advances. Anal Chem. 2017;89:430–452.

[5]   Tang M, Chan CK, Li YJ, Su H, Ma Q, Wu Z, et al. A review of experimental techniques for aerosol hygroscopicity studies. Atmos Chem Phys. 2019;19:12631–12686.

[6]   Hiranuma N, Augustin-Bauditz S, Bingemer H, Budke C, Curtius J, Danielczok A, et al. A comprehensive laboratory study on the immersion freezing behavior of illite NX particles a comparison of 17 ice nucleation measurement techniques. Atmos Chem Phys. 2015;15:2489–2518.

[7]   Garimella S, Kristensen TB, Ignatius K, Welti A, Voigtlander J, Kulkarni GR, et al. The spectrometer for ice nuclei (SPIN): An instrument to investigate ice nucleation. Atmos Chem Phys. 2016;9:2781–2795.

[8]   Riemer N, Ault AP, West M, Craig RL, Curtis JH. Aerosol mixing state: Measurements, modeling, and impacts. Rev Geophys. 2019;57(2):187–249.

[9]   Laskin A, Moffet RC, Gilles MK. Chemical imaging of atmospheric particles. Acc Chem Res. 2019;52(12):3419–3431.

[10]   Laskin A, Laskin J, Nizkorodov SA. Chemistry of atmospheric brown carbon. Chem Rev. 2015;115(10):4335–4382.

[11]   Reid JP, Bertram AK, Topping DO, Laskin A, Martin ST, Petters MD, et al. The viscosity of atmospherically relevant organic particles. Nature Comm. 2018;9(956):1–14.

[12]   You Y, Smith ML, Song M, Martin ST, Bertram AK. Liquid–liquid phase separation in atmospherically relevant particles consisting of organic species and inorganic salts. Int Rev Phys Chem. 2014;33(1):43–77.

[13]   Pye HOT, Nenes A, Alexander B, Ault AP, Barth MC, Clegg SL, et al. The acidity of atmospheric particles and clouds. Atmos Chem Phys. 2020;20:4809–4888.

[14]   Freedman MA, Ott E-JE, Marak KE. Role of pH in aerosol processes and measurement challenges. J Phys Chem A. 2019;123:1275–1284.

[15]   Hallquist M, Wenger JC, Baltensperger U, Rudich Y, Simpson D, Claeys M, et al. The formation, properties and impact of secondary organic aerosol: Current and emerging issues. Atmos Chem Phys. 2009;9:5155–5236.

[16]   Johnston MV, Kerecman DE. Molecular characterization of atmospheric organic aerosol by mass spectrometry. Ann Rev Anal Chem. 2019;12:247–274.

[17]   Duarte RMBO, Duarte AC. NMR Studies of Organic Aerosols. In: Webb GA, editor. Ann Rep NMR Spec. 922017. p. 83–135.

[18]   Lee S-H, Allen HC. Analytical measurements of atmospheric urban aerosol. Anal Chem. 2012;84:1196–1201.

[19]   Prather KM, Hatch CD, Grassian VH. Analysis of atmospheric aerosols. Annu Rev Anal Chem. 2008;1:485–514.

[20]   Solomon PA, Fraser MP, Herckes P. Methods for Chemical Analysis of Atmospheric Aerosols. In: Kulkarni P, Baron PA, Willeke K, editors. Aerosol Measurement: Principles, Techniques, and Applications. 3rd edn, Hoboken, NJ: John Wiley & Sons, Inc.; 2011. p. 153–177.

[21]   Wu L, Ro C-U. Aerosol hygroscopicity on a single particle level using microscopic and spectroscopic techniques: A review. Asian J Atmos Environ. 2020;14(3):177–209.

[22]   Physical Chemistry of Environmental Interfaces. In: Fairbrother H, Geiger FM, Grassian VH, Hemminger JC editors. J Phys Chem C. 113, Issue 622009. p. 2035–2646.

[23]   Physical Chemistry of Aerosols. In: Signorell R, Bertram AK, editors. Phys Chem Chem Phys. 112009. p. 7741–8104.

[24]   Finlayson-Pitts BJ. Reactions at surfaces in the atmosphere: Integration of experiments and theory as necessary (but not necessarily sufficient) for predicting the physical chemistry of aerosols. Phys Chem Chem Phys. 2009;11(36):7760–7779.

[25]   Gerber RB, Varner ME, Hammerich AD, Riikonen S, Murdachaew G, Shemesh D, et al. Computational studies of atmospherically-relevant chemical reactions in water clusters and on liquid water and ice surfaces. Acc Chem Res. 2015;48:399–406.

[26]   Grassian VH. Surface science of complex environmental interfaces: Oxide and carbonate surfaces in dynamic equilibrium with water vapor. Surf Sci. 2008;602:2955–2962.

[27]   Starr DE, Liu Z, Havecker M, Knop-Gericke A, Bluhm H. Investigation of solid/vapor interfaces using ambient pressure X-ray photoelectron spectroscopy. Chem Soc Rev. 2013;42:5833–5857.

[28] Kanji ZA, Ladino LA, Wex H, Boose Y, Burkert-Kohn M, Cziczo DJ, et al. Overview of ice nucleating particles. Meteorol Monogr. 2017;58:1–33.

[29] Li W, Sun J, Xu L, Shi Z, Riemer N, Sun Y, et al. A conceptual framework for mixing structures in individual aerosol particles. J Geophys Res-Atmos. 2016;121:13784–13798, doi:10.1002/2016JD025252.

[30] Song M, Marcolli C, Krieger UK, Lienhard DM, Peter T. Morphologies of mixed organic/inorganic/aqueous aerosol droplets. Faraday Discuss. 2013;165(1):289–316.

[31] Fundamentals of viscosity. [Available from: https://resources.saylor.org/wwwresources/archived/site/wp-content/uploads/2011/04/Viscosity.pdf.

[32] Koop T, Bookhold J, Shiraiwa M, Pöschl U. Glass transition and phase state of organic compounds: Dependency on molecular properties and implications for secondary organic aerosols in the atmosphere. Phys Chem Chem Phys. 2011;13(43):19238–19255.

[33] Shiraiwa M, Seinfeld JH. Equilibration timescale of atmospheric secondary organic aerosol partitioning. Geophys Res Lett. 2012;39:L24801.

[34] Shiraiwa M, Li Y, Tsimpidi AP, Karydis VA, Berkemeier T, Pandis SN, et al. Global distribution of particle phase state in atmospheric secondary organic aerosols. Nature Comm. 2017;8:1–7.

[35] Maclean AM, Butenhoff CL, Grayson JW, Barsanti K, Jimenez JL, Bertram AK. Mixing times of organic molecules within secondary organic aerosol particles: A global planetary boundary layer perspective. Atmos Chem Phys. 2017;17:13037–13048.

[36] Jimenez JL, Canagaratna MR, Donahue NM, Prevot ASH, Zhang Q, Kroll JH, et al. Evolution of organic aerosols in the atmosphere. Science. 2009;326(5959):1525–1529.

[37] Pankow JF. An absorption model of the gas/aerosol partitioning involved in the formation of secondary organic aerosol. Atmos Environ. 1994;28:189–193.

[38] O'Brian RE, Neu A, Epstein SA, MacMillan AC, Wang B, Kelley ST, et al. Physical properties of ambient and laboratory-generated secondary organic aerosol. Geophys Res Lett. 2014;41:4347–4353, doi:10.1002/2014GL060219.

[39] Pajunoja A, Malila J, Hao L, Joutsensaari J, Lehtinen KE, Virtanen A. Estimating the viscosity range of SOA particles based on their coalescence time. Aerosol Sci Technol. 2014;48:i–iv.

[40] Zhang Y, Sanchez MS, Douet C, Wang Y, Bateman AP, Gong Z, et al. Changing shapes and implied viscosities of suspended submicron particles. Atmos Chem Phys. 2015;15:7819–7829.

[41] Hosny NA, Fitzgerald C, Vysniauskas A, Athanasiadis A, Berkemeier T, Uygur N, et al. Direct imaging of changes in aerosol particle viscosity upon hydration and chemical aging. Chem Sci. 2016;7:1357–1367.

[42] Jarvinen E, Ignatius K, Nichman L, Kirstensen TB, Fuchs C, Hoyle CR, et al. Observation of viscosity transition in a-pinene secondary organic aerosol. Atmos Chem Phys. 2016;16:4423–4438.

[43] Power RM, Simpson SH, Reid JP, Hudson AJ. The transition from liquid to solid-like behaviour in ultrahigh viscosity aerosol particles. Chem Sci. 2013;4:2597–2604.

[44] Renbaum-Wolff L, Grayson JW, Bateman AP, Kuwata M, Sellier M, Murray BJ, et al. Viscosity of a-pinene secondary organic material and implications for particle growth and reactivity. PNAS. 2013;110:8014–8019.

[45] Renbaum-Wolff L, Grayson JW, Bertram AK. Technical note: New methodology for measuring viscosities in small volumes characteristic of environmental chamber particle samples. Atmos Chem Phys. 2013;13:791–802.

[46] Virtanen A, Joutsensaari J, Koop T, Kannosto J, Yli-Pirila P, Leskinen J, et al. An amorphous solid state of biogenic secondary organic aerosol particles. Nature. 2010;467:824–827.

[47] Bateman AP, Belassein H, Martin ST. Impactor apparatus for the study of particle rebound: Relative humidity and capillary forces. Aerosol Sci Technol. 2014;48:42–52.

[48] Bond T, Bergstrom R. Light absorption by carbonaceous particles: An investigative review. Aerosol Sci Technol. 2006;40(1):27–67.

[49] Laskina O, Young MA, Kleiber PD, Grassian VH. Infrared extinction spectra of mineral dust aerosol: Single components and complex mixtures. J Geophys Res. 2012;117:D18210, doi:10.1029/2012JD017756.

[50] Lack DA, Cappa CD. Impact of brown and clear carbon on light absorption enhancement, single scatter albedo and absorption wavelength dependence of black carbon. Atmos Chem Phys. 2010;10:4207–4220.

[51] Cappa CD, Lack DA, Burkholder JB, Ravishankara A. Bias in filter-based aerosol light absorption measurements due to organic aerosol loading: evidence from laboratory measurements. Aerosol Sci Technol. 2008;42:1022–1032.

[52] Subramanian R, Roden CA, Boparai P, Bond TC. Yellow beads and missing particles: Trouble ahead for filter-based absorption measurements. Aerosol Sci Technol. 2007;41:630–637.

[53] Ma L, Thompson JE. Optical properties of dispersed aerosols in the near ultraviolet (355 nm): measurement approach and initial data. Anal Chem. 2012;84(13):5611–5617.

[54] Dial KD, Hiemstra S, Thompson JE. Simultaneous measurement of optical scattering and extinction on dispersed aerosol samples. Anal Chem. 2010;82:788–7896.

[55] Hennigan CJ, Izumi J, Sullivan AP, Weber RJ, Nenes A. A critical evaluation of proxy methods used to estimate the acidity of atmospheric particles. Atmos Chem Phys. 2015;15(5):2775–2790.

[56] Li J, Jang M. Aerosol acidity measurement using colorimetry coupled with a reflectance UV-visible spectrometer. Aerosol Sci Technol. 2012;46:833–842.

[57] Craig RL, Peterson PK, Nandy L, Lei Z, Hossian MA, Camarena S, et al. Direct determination of aerosol pH: Size-resolved measurements of submicrometer and supermicrometer aqueous particles. Anal Chem. 2018;90:11232–11239.

[58] Ganor E. A method for identifying sulfuric acid in single cloud and fog droplets. Atmos Environ. 1999;33:4235–4242.

[59] Coddens EM, Angle KJ, Grassian VH. Titration of aerosol pH through droplet coalescence. J Phys Chem Lett. 2019;10:4476–4483.

[60] Buck RP, Rondinini S, Covington AK, Baucke FGK, Brett Christopher MA, Camoes MF, et al. Measurement of pH. Definition, standards, and procedures (IUPAC Recommendations 2002). Pure Appl Chem. 2002;11:2169–2200.

[61] Zhang Q, Jimemez JL, Canagaratna MR, Ulbrich IM, Ng N, K, Worsnop DR, et al. Understanding atmospheric organic aerosols via factor analysis of aerosol mass spectrometry: A review. Anal Bioanal Chem. 2011;401:3045–3067.

[62] Lin P, Laskin J, Nizkorodov SA, Laskin A. Revealing brown carbon chromophores produced in reactions of methylglyoxal with ammonium sulfate. Environ Sci Technol. 2015;49(24):14257–14266.

[63] Lin P, Liu J, Shilling JE, Kathmann SM, Laskin J, Laskin A. Molecular characterization of brown carbon (BrC) chromophores in secondary organic aerosol generated from photo-oxidation of toluene. Phys Chem Chem Phys. 2015;17(36):23312–23325.

[64] Timonen H, Aurela M, Carbone S, Saarnio K, Saarikoski S, Makela T, et al. High time-resolution chemical characterization of the water-soluble fraction of ambient aerosols with PILS-TOC-IC and AMS. Atmos Meas Tech. 2010;3:1063–1074.

[65] Saarikoski S, Sillanpaa M, Sofiev M, Timonen H, Saarnio K, Teinila K, et al. Chemical composition of aerosols during a major biomass burning episode over northern Europe in spring 2006: Experimental and modelling assessments. Atmos Environ. 2007;41:3577–3589.

[66]   Reggente M, Dillner AM, Takahama S. Analysis of functional groups in atmospheric aerosols by infrared spectroscopy: Systematic intercomparison of calibration methods for US measurement network samples. Atmos Meas Tech. 2019;12:2287–2312.

[67]   Cao G, Yan Y, Zou X, Zhu R, Ouyang F. Applications of infrared spectroscopy in analysis of organic aerosols. Spectral Anal Rev. 2018;6:12–32.

[68]   Maria SF, Russell LM, Turpin BJ, Porcja RJ, Campos TL, Weber RJ, et al. Source signatures of carbon monoxide and organic functional groups in Asian pacific regional aerosol characterization experiment (ACE-Asia) submicron aerosol types. J Geophys Res. 2003;108:8637, doi:10.1029/2003JD003703, 2003.

[69]   Maria SF, Russell LM, Turpin BJ, Porcja RJ. FTIR measurements of functional groups and organic mass in aerosol samples over the Caribbean. Atmos Environ. 2002;36:5185–5196.

[70]   Coury C, Dillner AM. ATR-FTIR characterization of organic functional groups and inorganic ions in ambient aerosols at a rural site. Atmos Environ. 2009;43:940–948.

[71]   Coury C, Dillner AM. A method to quantify organic functional groups and inorganic compounds in ambient aerosols using attenuated total reflectance FTIR spectroscopy and multivariate chemometric techniques. Atmos Environ. 2008;42:5923–5932.

[72]   Duarte RMBO, Duarte AC. Unraveling the structural features of organic aerosols by NMR spectroscopy: A review. Magn Reson Chem. 2014;53:658–666.

[73]   Zhang R, Wang G, Guo S, Zamora ML, Ying Q, Lin Y, et al. Formation of urban fine particulate matter. Chem Rev. 2015;115:3803–3855.

[74]   Duncan MW. Quantitative Analysis by Mass Spectrometry: Some Important Considerations. In: Caprioli RM, Malorni A, Sindona G, editors. Selected Topics in Mass Spectrometry in the Biomolecular Sciences NATO ASI Series (Series C: Mathematical and Physical Sciences). vol. 504, Dordrecht: Sprigner; 1997. p. 103–119.

[75]   Marple VA, Rubow KL, Behm SM. A microorifice uniform deposit impactor (MOUDI): Description, calibration, and use. Aerosol Sci Technol. 1991;14(4):434–446.

[76]   Sardar SB, Fine PM, Mayo PR, Sioutas C. Size-fractionated measurements of ambient ultrafine particle chemical composition in Los Angeles using the nanoMOUDI. Environ Sci Technol. 2005;39(4):932–944.

[77]   Mainey A, Winberry WT Jr. Method IO-3.1: Selection, Preparation and Extraction of Filter Material. Cincinnati, OH: U.S. Environmental Protection Agency; 1999.

[78]   Meskhidze N, Volker C, Al-Abadleh HA, Barbeau K, Bressac M, Buck C, et al. Perspective on identifying and characterizing the processes controlling iron speciation and residence time at the atmosphere-ocean interface. Mar Chem. 2019;217(1–16):103704.

[79]   Cutter G, Casciotti K, Croot P, Geibert W, Heimbürger L-E, Lohan M, et al. Sampling and sample-handling protocols for GEOTRACES Cruises. Toulouse, France: GEOTRACES International Project Office; Version 3, August 2017, http://dx.doi.org/10.25607/OBP-2.

[80]   Usher CR, Michel AE, Grassian VH. Reactions on mineral dust. Chem Rev. 2003;103:4883–4940.

[81]   Jeong GY. Mineralogy and geochemistry of 5 Asian dust: Dependence on migration path, fractionation, and reactions with polluted air. Atmos Chem Phys. 2020;20:7411–7428.

[82]   Bluhm H, Siegmann HC. Surface science with aerosols. Surf Sci (Review). 2009;603:1969–1978.

[83]   Al-Abadleh HA, Grassian VH. Oxide surfaces as environmental interfaces. Surf Sci Rep. 2003;52:63–162.

[84]   Voges AB, Al-Abadleh HA, Geiger FM. Applications of Nonlinear Optical Techniques for Studying Heterogeneous Systems Relevant in the Natural Environment. In: Grassian VH, editor. Environmental Catalysis. Boca Raton, FL: CRC Press; 2005. p. 83–128.

[85] Budke C, Koop T. BINARY: An optical freezing array for assessing temperature and time dependence of heterogeneous ice nucleation. Atmos Meas Tech. 2015;8:689–703.

[86] Hill TCJ, Moffett BF, DeMott PJ, Georgakopoulos DG, Stump WL, Franc GD. Measurement of ice nucleation-active bacteria on plants and in precipitation by quantitative PCR. App Environ Microbiol. 2014;80(1256–1267).

[87] O'Sullivan D, Murray BJ, Malkin TL, Whale TF, Umo NS, Atkinson JD, et al. Ice nucleation by fertile soil dusts: Relative importance of mineral and biogenic components. Atmos Chem Phys. 2014;14:1853–1867.

[88] Diehl K, Debertshäuser M, Eppers O, Schmithüsen H, Mitra SK, Borrmann S. Particle surface area dependence of mineral dust in immersion freezing mode: Investigations with freely suspended drops in an acoustic levitator and a vertical wind tunnel. Atmos Chem Phys. 2014;14:12343–12355.

[89] Szakáll M, Diehl K, Mitra SK, Borrmann S. A wind tunnel study on the shape, oscillation, and internal circulation of large raindrops with sizes between 2.5 and 7.5 mm. J Atmos Sci. 2009;66:755–765.

[90] Diehl K, Mitra SK, Szakáll M, Blohn N, Borrmann S, Pruppacher HR. Chapter 2. Wind Tunnels: Aerodynamics, Models, and Experiments. In: Pereira JD, editor. The Mainz VerticalWind Tunnel Facility: A Review of 25 Years of Laboratory Experiments on Cloud Physics and Chemistry. Hauppauge, NY, USA: Nova Science Publishers, Inc.; 2011.

[91] Wright TP, Petters MD. The role of time in heterogeneous freezing nucleation. J Geophys Res Atmos. 2013;118:3731–3743.

[92] Schill GP, Tolbert MA. Heterogeneous ice nucleation on phase-separated organic-sulfate particles: Effect of liquid vs. Glassy Coat Atmos Chem Phys. 2013;13:4681–4695.

[93] Bingemer H, Klein H, Ebert M, Haunold W, Bundke U, Herrmann T, et al. Atmospheric ice nuclei in the Eyjafjallajökull volcanic ash plume. Atmos Chem Phys. 2012;12:857–867.

[94] Möhler O, Stetzer O, Schaefers S, Linke C, Schnaiter M, Tiede R, et al. Experimental investigation of homogeneous freezing of sulphuric acid particles in the aerosol chamber AIDA. Atmos Chem Phys. 2003;3:211–223.

[95] Hiranuma N, Paukert M, Steinke I, Zhang K, Kulkarni G, Hoose C, et al. A comprehensive parameterization of heterogeneous ice nucleation of dust surrogate: Laboratory study with hematite particles and its application to atmospheric models. Atmos Chem Phys. 2014;14:13145–13158.

[96] Hiranuma N, Hoffmann N, Kiselev A, Dreyer A, Zhang K, Kulkarni G, et al. Influence of surface morphology on the immersion mode ice nucleation efficiency of hematite particles. Atmos Chem Phys. 2014;14:2315–2324.

[97] Tobo Y, Prenni AJ, DeMott PJ, Huffman JA, McCluskey CS, Tian G, et al. Biological aerosol particles as a key determinant of ice nuclei populations in a forest ecosystem. J Geophys Res-Atmos. 2013;118:10100–10110.

[98] Hoffmann N, Kiselev A, Rzesanke D, Duft D, Leisner T. Experimental quantification of contact freezing in an electrodynamic balance. Atmos Meas Tech. 2013;6:2373–2382.

[99] Bundke U, Nillius B, Jaenicke R, Wetter T, Klein H, Bingemer H. The fast ice nucleus chamber FINCH. Atmos Res. 2008;90:180–186.

[100] Hartmann S, Niedermeier D, Voigtländer J, Clauss T, Shaw RA, Wex H, et al. Homogeneous and heterogeneous ice nucleation at LACIS: Operating principle and theoretical studies. Atmos Chem Phys. 2011;11:1753–1767.

[101] Wex H, DeMott PJ, Tobo Y, Hartmann S, Rösch M, Clauss T, et al. Kaolinite particles as ice nuclei: Learning from the use of different kaolinite samples and different coatings. Atmos Chem Phys. 2014;14:5529–5546.

[102] Tajiri T, Yamashita K, Murakami M, Orikasa N, Saito A, Kusunoki K, et al. A novel adiabatic-expansion-type cloud simulation chamber. J Meteor Soc Japan. 2013;91:687–704.

[103] Chou C, Stetzer O, Weingartner E, Juranyi Z, Kanji ZA, Lohmann U. Ice nuclei properties within a Saharan dust event at the Jungfraujoch in the Swiss Alps. Atmos Chem Phys. 2011;11:4725–4738.

[104] Kanji ZA, Welti A, Chou C, Stetzer O, Lohmann U. Laboratory studies of immersion and deposition mode ice nucleation of ozone aged mineral dust particles. Atmos Chem Phys. 2009;13:9097–9118.

[105] Friedman B, Kulkarni G, Beránek J, Zelenyuk A, Thornton JA, Cziczo DJ. Ice nucleation and droplet formation by bare and coated soot particles. J Geophys Res. 2011;116:D17203, doi:10.1029/2011JD015999.

[106] Lüönd F, Stetzer O, Welti A, Lohmann U. Experimental study on the ice nucleation ability of size-selected kaolinite particles in the immersion mode. J Geophys Res. 2010;115:D14201, doi:10.1029/2009JD012959.

[107] Stetzer O, Baschek B, Luond F, Lohmann U. The Zurich ice nucleation chamber (ZINC) – A new instrument to investigate atmospheric ice formation. Aerosol Sci Technol. 2008;42:64–74.

[108] Welti A, Lüönd F, Stetzer O, Lohmann U. Influence of particle size on the ice nucleating ability of mineral dusts. Atmos Chem Phys. 2009;9:6705–6715.

[109] Bateman AP, Bertram AK, Martin ST. Hygroscopic influence on the semisolid-to-liquid transition of secondary organic materials. J Phys Chem A. 2015;119:4386–4395.

[110] Dallemagne MA, Huang XY, Eddingsaas NC. Variation in pH of model secondary organic aerosol during liquid–liquid phase separation. J Phys Chem A. 2016;120(18):2868–2876.

[111] Craig RL, Nandy L, Axson JL, Dutcher CS, Ault AP. Spectroscopic determination of aerosol pH from acid–base equilibria in inorganic, organic, and mixed systems. J Phys Chem A. 2017;121(30):5690–5699.

[112] Rindelaub JD, Craig RL, Nandy L, Bondy AL, Dutcher CS, Shepson PB, et al. Direct measurement of pH in individual particles via Raman microspectroscopy and variation in acidity with relative humidity. J Phys Chem A. 2016;120:911–917.

[113] Bondy AL, Craig RL, Zhang Z, Gold A, Surratt JD, Ault AP. Isoprene-derived organosulfates: Vibrational mode analysis by Raman spectroscopy, acidity-dependent spectral modes, and observation in individual atmospheric particles. J Phys Chem A. 2018;122:303–315.

[114] Wei H, Vejerano EP, Leng W, Huang Q, Willner MR, Marr LC, et al. Aerosol microdroplets exhibit a stable pH gradient. PNAS. 2018;115(28):7272–7277.

[115] Colussi AJ. Can the PH at the air/water interface be different from the pH of bulk water?. PNAS. 2018;115:E7887.

[116] Cohen L, Quant MI, Donaldson DJ Real-time measurements of pH changes in single, acoustically levitated droplets due to atmospheric multiphase chemistry. ACS Earth Space Chem. 2020;4(6):854–861.

[117] Decesari S, Mircea M, Cavalli F, Fuzzi S, Moretti F, Tagliavini E, et al. Source attribution of water-soluble organic aerosol by nuclear magnetic resonance spectroscopy. Environ Sci Technol. 2007;41:2479–2484.

[118] Bilo F, Borgese L, Wambui A, Assi A, Zacco A, Federici S, et al. Comparison of multiple X-ray fluorescence techniques for elemental analysis of particulate matter collected on air filters. J Aerosol Sci. 2019;122:1–10.

[119] Zhou X, Liu D, Bu H, Deng L, Liu H, PY, et al. XRD-based quantitative analysis of clay minerals using reference intensity ratios, mineral intensity factors, Rietveld, and full pattern summation methods: A critical review. Solid Earth Sci. 2018;3:16–29.

[120] Pröfrock D, Prange A. Inductively coupled plasma-mass spectrometry (ICP-MS) for quantitative analysis in environmental and life sciences: A review of challenges, solutions, and trends. Appl Spectrosc. 2012;66(8):843–868.

[121] Method 200.2, Revision 2.8: Sample Preparation Procedure for Spectrochemical Determination of Total Recoverable Elements. https://19january2017snapshot.epa.gov/sites/production/files/5-08/documents/method_200-2_rev_2-8_1994.pdf.

Chapter 3
Physical properties of aerosols

Physical properties of atmospheric aerosol particles are those that can be measured without changing the chemical composition. They govern interactions with solar radiation and clouds that explain and quantify the direct and indirect effects of aerosols on the climate system. These properties change during the atmospheric lifetime of the aerosol particles due to aging processes that alter their chemical composition. Table 1.2 in Chapter 1 lists the definition of these properties. They include aerosol size and mass concentration of field-collected primary and secondary atmospheric aerosol particles listed in Table 1.1, hygroscopic properties and ice nucleation, aerosol morphology and mixing states, optical properties, viscosity, liquid–liquid phase separation, and acidity. In chapter 2, recent advances in the analytical tools capable of measuring physical properties of aerosols are summarized. In this chapter, specific examples are presented on each of these properties from the synthesis of relevant literature.

3.1 Hygroscopic properties

The term "atmospheric liquid water (ALW)" refers to the water content (i.e., water activity) in the condensed phase of atmospheric particles, droplets, and surfaces. ALW varies with temperature, relative humidity (RH), particle size, and surface tension sensitive to chemical composition, which, in turn, controls water solubility, viscosity, and hydrophilicity. Hence, quantifying gas phase water uptake on atmospheric particles leads to understanding their role in air quality, climate change [1], and health [2]. Figure 3.1 shows a general schematic of the effect of RH and temperature on particle phase transitions, surface versus bulk chemistry, formation of haze, cloud condensation nuclei (CCN), and ice nuclei (IN) [1].

In thermodynamic terms, the activation of particles to CCN and IN is briefly described as follows: water activity, a_w, is the effective concentration of water that accounts for nonideal chemical interactions in solution, and its value in atmospheric particles and droplets is related to the actual vapor pressure of gas phase water, p_w, over these systems per the following equation [3]:

$$a_w = \frac{p_w}{p_{w,sat}(T)} \tag{3.1}$$

where $p_{w,sat}(T)$ is the temperature-dependent saturation vapor pressure of water. The value of a_w is related to the ambient RH through the following equation:

https://doi.org/10.1515/9781501519376-003

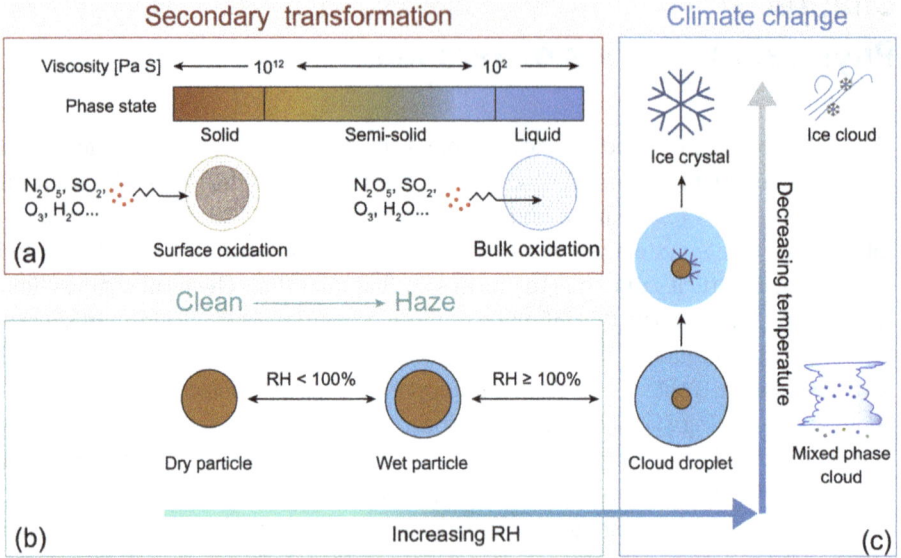

Figure 3.1: Interaction of water with atmospheric particles: (a) particle phase transition from solid to liquid, with increasing RH; (b) particles taking up water leads to an increase in particle size and mass loading; (c) particles participate in cloud formation by serving as CCN and IN. Reproduced from reference [1] with permission under the Creative Commons CC-BY license, © Oxford University Press, 2018.

$$RH = \frac{p_{w,amb}}{p_{w,sat}(T)} \times 100 = a_w exp\left(\frac{4\sigma_{obs}M_w}{RT\rho_w D_{p,wet}}\right) \qquad (3.2)$$

where $p_{w,amb}$ is the ambient (i.e., observed) vapor pressure of water, σ_s is the surface tension of the wet particle at the solution/air interface, which in the case of pure water, $\sigma_{obs} = \sigma_w = 72.8 \times 10^{-3}$ N m^{-1} (72.8 dyn cm^{-1}) at 293 K, M_w is the molecular weight of water, R is the ideal gas constant, T is temperature (in degrees Kelvin), ρ_w is the density of water, and $D_{p,wet}$ is the (spherical) diameter of the wet particle (or droplet, for particles that have sufficient water content) [3]. The exponential term is called the Kelvin effect term which takes into account the fact that the vapor pressure over a surface increases with surface curvature, over an "infinitely flat" surface with a diameter greater than 1 μm at equilibrium, RH ~ a_w. For particles containing surface-active compounds, mostly organics, the observed surface tension of water is lowered, which tends to follow the Szyskowski–Langmuir adsorption isotherm:

$$\sigma_{obs} = \sigma_w - RTT_\infty \ln\left(1 + \frac{C}{a_L}\right) \qquad (3.3)$$

where Γ_∞ is the maximum surface excess (mol m^{-2}), c is the bulk concentration of a surface-active species (mol m^{-3}), and a_L is the bulk concentration at half surface coverage (mol m^{-3}).

Under non equilibrium conditions for RH values greater than 100%, the RH is expressed as percent supersaturation (s) [3]:

$$s(\%) = \left(\frac{RH}{100} - 1\right) \times 100 \tag{3.4}$$

When RH or s is plotted as a function of $D_{p,wet}$, the resulting curve is the Köhler curve, which is shown in Figure 3.2, for different aerosol materials of various hygroscopicities. The maximum in the particle vapor pressure (or %s) is called the critical supersaturation (s_{crit}), which occurs at a particular $D_{p,wet}$. Hence, the term "particle activation" refers to the nonequilibrium conditions under which a particle size reaches the size corresponding to s_{crit}, where it gets activated into a cloud droplet and is able to grow spontaneously. In cloud systems, the values of maximum s range from 0.1–1% [3].

The following sections describe, in detail, the qualitative and quantitative aspects of ALW on aerosol components with various solubilities, to help understand hygroscopic properties of field-collected aerosol particles.

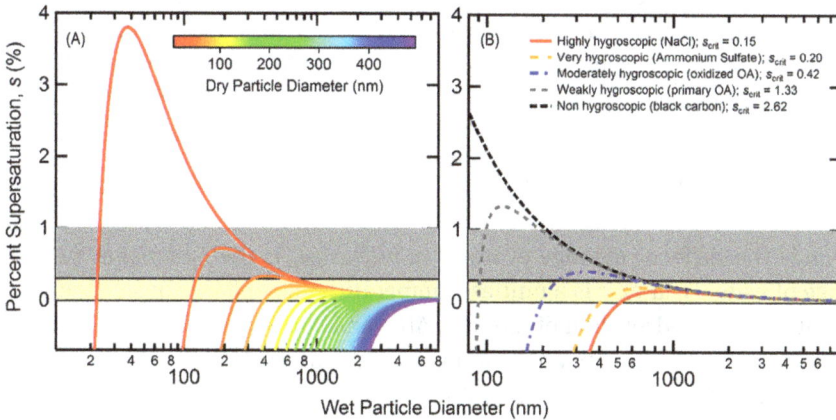

Figure 3.2: (A) Köhler curves for ammonium sulfate particles with dry particle diameters ranging from 10 to 490 nm, in 20 nm steps. The line color indicates the particle dry diameter. The gray-filled region indicates where s < 1.0% and the yellow-filled region, where s < 0.3%. Note the decrease in the critical supersaturation with increasing particle size. (B) Köhler curves for particles with 80 nm dry diameters but differing composition. Particles have been chosen to represent highly hygroscopic (NaCl), very hygroscopic (ammonium sulfate), moderately hygroscopic (oxidized organic aerosol), weakly hygroscopic (primary organic aerosol), and nonhygroscopic (black carbon) materials. Reprinted (adapted) with permission from reference [3], © American Chemical Society, 2020.

3.1.1 Gas phase water uptake on insoluble aerosol components

Examples of insoluble aerosol particles are freshly emitted mineral dust and hydro-phobic organics. Uptake of gas phase water on the surfaces of these materials re-sults in the formation of "adsorbed water", which can take the form of either thin films or islands, depending on the thermodynamic favorability of hydrogen bond-ing with the underlying surface. The number of hydrogen bonding donors and ac-ceptors is a strong indicator of the extent of water–surface interactions.

3.1.1.1 Metal oxides, clays, and carbonates

Mineral dust particles contain metal oxides and clay minerals, reflective of the chemical composition of their source regions. The top three oxides in the Earth's crust are SiO_2, Al_2O_3, and Fe_2O_3 [4], and the top three clay minerals by atmospheric loadings are illite, montmorillonite, and kaolinite [5]. The surfaces of metal oxides in mineral dust particles are terminated with hydroxyl groups that serve as hydro-gen bonding donors and acceptors [6]. Each metal oxide is unique in the type and surface density of hydroxyl groups, and hence, mineralogy of dust plays a role in the extent of its water uptake. Also, the water-soluble inorganic content plays a role in water uptake on dust samples [7, 8]. Gas phase water uptake as a function of RH is typically described using thermodynamic isotherm models that describe multi-layer formation of adsorbed water. A three-parameter Brunauer–Emmett–Teller (BET) model, described by the following equation, was found to reproduce the ex-perimental data quite well on metal oxides [9]:

$$m = \left[\frac{m_{ML}c(RH)}{1-(RH)}\right]\left[\frac{1-(n+1)(RH)^n + n(RH)^{n+1}}{1+(c-1)(RH) - c(RH)^{n+1}}\right] \tag{3.5}$$

where m is the surface coverage of water (g m^{-2}), m_{ML} is the monolayer surface coverage of water (g m^{-2}), c is a unitless temperature-dependent constant related to the enthalpies of adsorption of the first, ΔH_1°, and subsequent layers (expressed as the standard enthalpy for water vapor condensation, $\Delta H_{cond}^\circ = -44$ kJ mol^{-1}), $c = exp\left(-\left[\Delta H_1^\circ - \Delta H_{cond}^\circ\right]/RT\right)$, n is a unitless fitting parameter that represents the maximum number of layers of adsorbed species and is related to the pore size and properties of the adsorbent, R is the gas constant, and T is temperature in K.

This model is derived for the formation of a finite number of layers, as opposed to the 2-fit parameter BET model [5, 10] derived for adsorption on uniform surfaces with infinite number of layers. As reviewed in Tang et al. [5] and Chen et al. [11], the RH for the formation of the first adsorbed water layer on SiO_2, TiO_2, MgO, α-Al_2O_3, and Fe_2O_3 particles ranges from 24–30% at 298 K. The adsorption enthalpies of water in the first layer are higher than the condensation enthalpy of water, -44 kJ mol^{-1}, suggesting stronger water–surface interactions than water–water interactions.

Gas phase water uptake on clays ubiquitous in mineral dust, such as kaolinite ([Si_4] $Al_4O_{10}(OH)_8$), montmorillonite ($M_x[Si_8]Al_{3.2}Fe_{0.2}Mg_{0.6}O_{20}(OH)_4$), and illite ($M_x[Si_{6.8}Al_{1.2}]$ $Al_3Fe_{0.25}Mg_{0.75}O_{20}(OH)_4$, where M = interlamellar cation, Li^+, K^+, Na^+, Ca^{2+}, or Mg^{2+} in the space of the clay layers) was also investigated using different techniques, as summarized in the review by Tang et al. [5]. To highlight the usefulness of adsorption models in quantifying adsorption strength, one of these studies will be described in detail, below, for comparison with water uptake on metal oxides described above. Clays are structurally different from pure metal oxides in that they are phyllosilicates composed of plate-like structures of alternating layers of alumina and silica sheets [12]. The ability of these clays to adsorb water molecules between their aluminosilicate layers can result in clay expansion, as observed in the case of montmorillonite, driven mostly by the hydration of exchangeable cations, M. Using spectral data from ATR-FTIR spectroscopy experiments designed to quantify the surface coverage of adsorbed water on these clays as a function of RH, Hatch et al. [12] found that the Freundlich adsorption model fits the data well over the entire range of RH values studied. Figure 3.3 shows the clay water content (expressed as mass of adsorbed water by clay mass, g_{H2O}/g_{clay}) as a function of %RH, with best fits, using the Freundlich adsorption model shown in the following equation [12]:

$$\frac{g_{H2O}}{g_{clay}} = k \left(\frac{P}{P_o} \right)^{1/n}$$

(3.6)

where P/P_o is the relative pressure of water vapor at temperature T (i.e., RH), and k and n are the Freundlich constants representing the adsorption capacity and adsorption strength, respectively. Strong adsorption is characterized by n values greater than 1. The adsorption curves revealed two distinct water adsorption regimes: below 20% where water-surface interactions are dominant, and above 20% where bulk water-water interactions are dominant in the multilayers. Values of the Freundlich model best fit parameters in the low RH region (i.e., below 20%RH) showed that all clays have a strong affinity for gas phase water adsorption and that montmorillonite ($n = 5.1$) has the highest water adsorption strength and highest adsorption capacity at RH values greater than 19%, relative to the kaolinite ($n = 1.7$) and illite ($n = 2.7$) clays. For the high RH region (i.e., above 20% RH), n values were less than 1. The difference in the hygroscopicity of montmorillonite compared to kaolinite and illite was explained by hydration heterogeneity and pore filling processes [12].

In the review by Tang et al. [5], the best fit thermodynamic parameters of water adsorption on a given material were used to calculate the surface coverage of water, for comparison among different experimental results that used the BET and Freundlich adsorption models. Results from hygroscopic growth experiments of aerosolized samples that measured change in particle diameter, D, with RH were modeled using the hygroscopic growth theory, where the single hygroscopicity parameter, κ, is related to the surface coverage of water, θ, through the following equation:

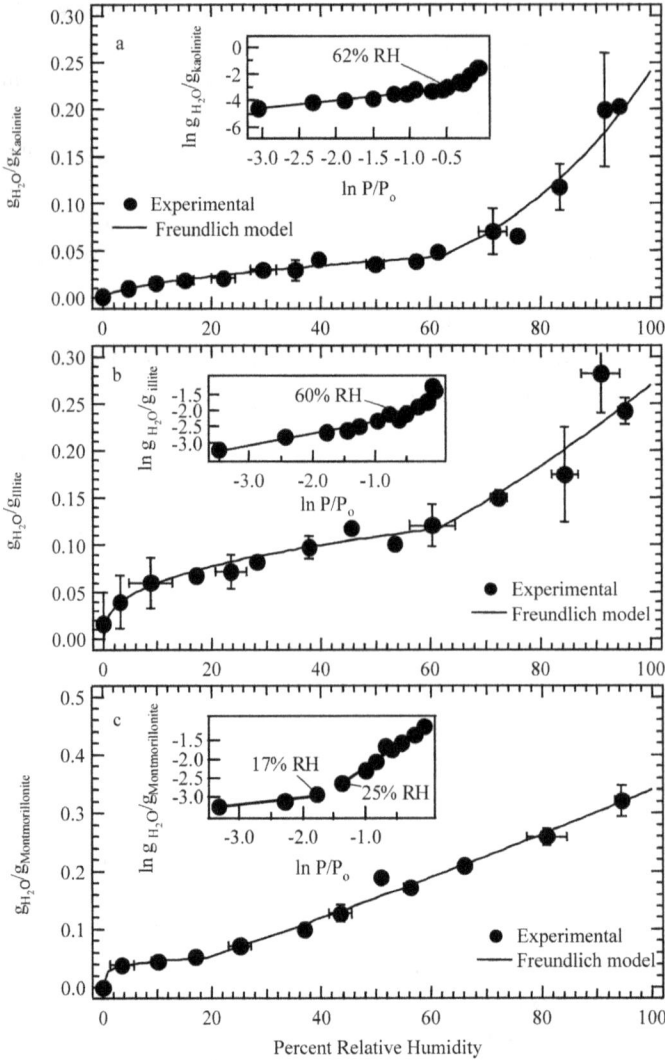

Figure 3.3: Measured and modeled (Freundlich) clay water content (g_{H2O}/g_{clay}) at 298 K as a function of RH from 0% to 94% RH for (a) kaolinite, (b) illite, and (c) montmorillonite clays. The scale on the y-axis varies. The insets represent the data fit to the linear form of the Freundlich adsorption isotherm. The ratio, g_{H2O}/g_{clay}, was calculated by converting the total number of adsorbed water molecules (from the ATR–FTIR absorbance of the bending mode of water at 1640 cm^{-1}) to a mass and dividing by the mass of the clay sample (see reference [12] for details). Reprinted (adapted) with permission from reference [12], © American Chemical Society, 2012.

$$\theta = \frac{D - D_d}{2D_w} = \frac{GF - 1}{2} \cdot \frac{D_d}{D_w}, \text{ where } GF = \sqrt{1 + \kappa \frac{RH}{1 - RH}} \tag{3.7}$$

where D and D_d are the diameters of the wet particle and the dry particle, respectively, D_w is the average diameter of a water molecule adsorbed on the particle surface (assumed to be 0.275 nm) [5], and GF is the growth factor defined as the ratio of the diameter of a particle at a given RH to that of the dry particle. In their paper, Mirrielees and Brooks provided a detailed analysis of the uncertainties associated with κ [13]. Figures 3.4a–c show a comparison of the simulated surface coverage for the most abundant metal oxides and clays in mineral dust. These simulations agree to some extent in the low- to medium-RH and show much larger variability in the high-RH region. The reasons behind this variability are sample source and preparation, which suggests the need for standardizing materials and procedures for quantifying hygroscopic properties.

Calcium carbonate ($CaCO_3$) in the calcite phase is also abundant in mineral dust, following quartz and major clay minerals. The hygroscopic properties of calcite received much attention because of its abundance and reactivity with acidic gases such as nitric acid. Tang et al. [5] reviewed the water uptake and CCN studies on fresh and reacted calcite, which reported using the water adsorption and hygroscopic growth models described above. Figure 3.4d shows the simulated surface water coverage curves on fresh calcite as a function of RH, based on the best-fit parameters to experimental data. Similar to metal oxides and clays, there is a good agreement among the cited studies, below 40% RH. Within the uncertainties associated with each model, there is a better agreement in the higher RH region compared to metal oxides and clays. The use of calcite single crystals provided insights into surface reconstruction in humid environments [14]. Using alternating current (AC) mode, Atomic Force Microscopy (AFM) height images combined with force measurements and phase imaging, Baltrusaitis and Grassian [14] found that an amorphous hydrate layer forms upon the exposure of $CaCO_3(10\bar{1}4)$ to 70% RH at 296 K, which serves as a substrate for the crystallization of another structurally different layer proposed to be vaterite, another polymorph of calcium carbonate. The AFM images also revealed that calcite is highly inhomogeneous, with regions that vary in water content.

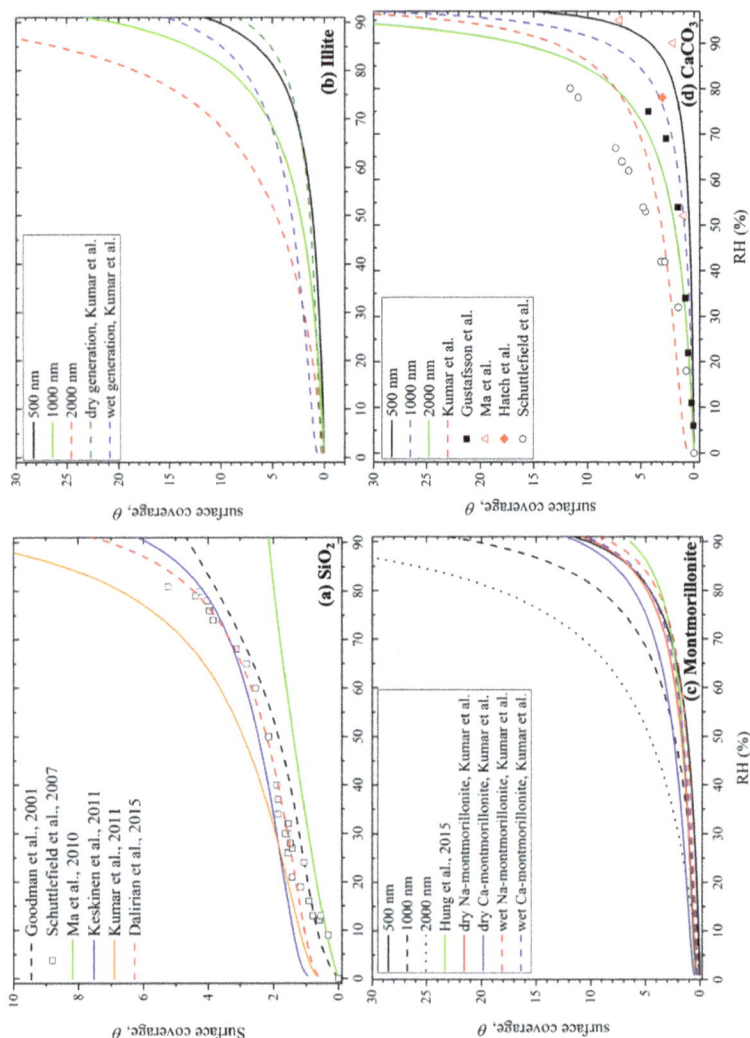

Figure 3.4: Comparison of simulated surface coverage of adsorbed water on (a) SiO$_2$, (b) illite, and (c) montmorillorite particles from the best-fit parameters of adsorption isotherm models and CCN activity measurements to experimental data reported in the following references: Goodman et al. [9], Schuttlefield et al. [15], Ma et al. [16], Keskinen et al. [17], Kumar et al. [18], Dalirian et al. [19], Hung et al. [20], Gustafsson et al. [21], and Kumar et al. [18, 22]. For (b) and (c), surface coverages of adsorbed water were calculated per the hygroscopic growth theory, using an average hygroscopicity parameter, κ, of 0.005 (see references [23, 24]) and assumed dry particle diameters of 500, 1000, and 2000 nm, respectively (see reference [5] for details). For (d), surface coverages of adsorbed water were calculated per the hygroscopic growth theory, using an average hygroscopicity parameter, κ, of 0.002 (see references [23, 24]) and assumed dry particle diameters of 500, 1000, and 2000 nm, respectively (see reference [5] for details). Reprinted (adapted) with permission from reference [5], © American Chemical Society, 2016.

3.1.1.2 Authentic mineral dust

The term "authentic mineral dust" is used to refer to dust samples collected from deserts around the world. Chen et al. [11] conducted hygroscopic experiments on seven authentic mineral dust samples from Africa, Asia, and North America. Table 3.1 lists the names of these dust samples, location, method of collection, BET surface area, average particle diameter, and mineralogy. The authors found that at 90%RH, the mass ratios of adsorbed water to the dry mineral ranged from 0.0011 to 0.3080, depending on the BET surface areas of mineral dust samples [11]. When the results were compared with those from other research groups, the authors found some degree of discrepancy in the hygroscopicity parameterizations, which were attributed to differences in samples and pre-treatment procedures used by each group. Hence, as concluded above, future work would benefit from distributing same samples among different groups and standardizing sample pre-treatment methods and for the water uptake data to be normalized to surface area of the material [5, 11].

Table 3.1: Characteristics of authentic dust samples used in the hygroscopicity studies by Chen et al. [11].

Name	Source region	Method of collection	BET surface area ($m^2\,g^{-1}$)	Average particle diameter, d_p (μm)	Mineralogy by X-ray diffraction XRD (%mass fraction)
Arizona test dust (ATD)	North America	Standard, from Powder Technology Inc.	36.67±1.06	1.05±0.20	Quartz (68–76), Aluminum oxide (10–15) Ca, Na, K, Fe oxides (each 2–5) [25]
China loess (SX dust)	China	Top soil from Shaanxi, Chinese Academy of Geological Science	11.71±0.02	2.44±0.42	Albite (44) Quartz (24), Muscovite (11), Orthoclase (10) [7]
QH dust	China	Top brown desert soil in Qinghai, Chinese Academy of Geological Science	8.79±0.02	18.56±2.38	Quartz (41), Albite (20), Muscovite (17), Orthoclase (11) [7]
TLF dust	China	Turpan, Xinjiang, China, urban, during dust storm	8.49±0.01	8.04±1.46	Not performed
Bordj dust	Africa	Top soil, Bordj, Algeria	16.40±1.20	32.30±3.06	Quartz (88), Aluminum oxide (4.4) [25]

Table 3.1 (continued)

Name	Source region	Method of collection	BET surface area (m² g⁻¹)	Average particle diameter, d_p (µm)	Mineralogy by X-ray diffraction XRD (%mass fraction)
M'Bour dust	Africa	Top soil, M'Bour, Senegal	14.50±1.00	54.41±5.99	Quartz (95), Aluminum oxide (1.2) [25]
Saharan dust (Rawdat)	Africa	Top soil, Cabo Verde	51.46±0.34	23.70±2.59	Quartz (61), Calcite (21), Albite (10), Aluminum oxide (4.4) [25]

Note: Albite = $NaAlSi_3O_8$, Muscovite = $KAl_2(Si_3AlO_{10})(OH)_2$, Orthoclase = $KAlSi_3O_8$.

Ibrahim et al. [26] studied water uptake on ATD as a function particle size (0–3, 5–10, 10–20, 20–40, and 40–80 µm) using diffuse reflectance infrared Fourier transform spectroscopy (DRIFTS) and thermogravimetric analysis at room temperature (295 K). The authors found that the RH for the formation of the water monolayer on the above samples increased with increasing particle size, at 13 ± 1, 17 ± 1, 22 ± 2, 25 ± 2, and 28 ± 2%, respectively, suggesting relative decrease in hygroscopic properties. Water desorption kinetics were found to follow a second-order model with apparent desorption rate coefficients that also increase with increasing particle size. The elemental analysis of these ATD samples shows that the Si content increases from 58 wt% to 83 wt%, and the Al and Fe content decreases from 17 wt% to 7 wt% and 6 wt% to 2 wt%, respectively, with increasing particle size. Other metals such as Ca, K, and Mg also showed a decreasing trend with particle size. These differences in the mineralogical composition with particle size affect the hydroxyl site density that plays a role in water uptake.

3.1.1.3 Anthropogenic mineral dust

As stated in Chapter 1, anthropogenic mineral dust refers to fugitive, combustion, and industrial dust (AFCID) that is mainly composed of metal oxides and aluminosilicates. Vu et al. [2] reviewed the variation in GF of particles generated from various sources, including nucleation, traffic emissions, and biomass burning, taking into account the spatial and temporal variations. They concluded that typical hygroscopicity of particles varies by source and air mass, shows a clear diurnal and seasonal trends with higher values found in daytime and summer, and increases during atmospheric aging, compared to hydrophobic and less hygroscopic freshly emitted traffic particles. The hygroscopicity of model AFCID particles was investigated by a number of groups, with the goal of correlating their hygroscopic growth with chemical composition. In the case of combustion particles, coal contains kaolinite and pyrite ($Fe_{5/4}S$), which, upon complete combustion, forms mullite ($3Al_2O_3 \cdot 2SiO_2$) and hematite (Fe_2O_3) [27, 28]. Using a vapor sorption analyzer and DRIFTS, Peng et al. [29] studied the hygroscopic properties of 11 AFCID samples that include oil and coal fly ash, road dust, and industrial dust. The majority of these samples were obtained as certified materials from commercial sources, used without pretreatment, and characterized for their soluble inorganic content and BET surface areas. All of the samples contained SO_4^{2-} and Mg^{2+}, half of them contained Na^+ and Ca^{2+}, and a quarter contained NH_4^+, K^+, and Cl^-. The samples were not analyzed for their iron content. Coal fly ash analyzed in other studies reported 4–9 wt% total iron, higher than that in ATD (~2 wt%) [30]. The BET surface area of the samples analyzed Peng et al. [29] studies ranges from 0.4 to 3 m^2 g^{-1}. Figure 3.5 shows the dependency of the mass ratio of adsorbed water relative to the dry mass of the samples on the water-soluble ion content and BET surface area. These data clearly show that samples with larger soluble ion contents (> 100 mg g^{-1}) and BET surface area (> 1.6 m^2 g^{-1}) have larger capacities to adsorb water at 90% RH.

In their study, Navea et al. [31, 32] analyzed coal fly ash samples from the US and Europe for their mineralogy and surface hydroxyl groups and found that mullite and hematite are major metal oxide end products from the combustion of coal. The BET surface area and mean diameter of the fly ashes ranged from 0.98–3.5 m^2 g^{-1} and 1.6–4.6 µm, respectively. Water uptake on these samples studied using ATR–FTIR and quartz crystal microbalance was best described using a Type II BET adsorption model at 295 K over the full RH range. The RH at which a monolayer of water formed ranged from 26–55%, which is higher than that observed for pure metal oxides and natural mineral dust [5, 11], suggesting lower water uptake for the coal fly ash samples. Although not directly measured, the authors attributed differences among fly ash and natural dust samples to differences in the water-soluble inorganic content and hydroxyl site density.

3.1.1.4 Water-insoluble organics

As stated in Chapter 1, organics from primary and secondary sources represent a major component of atmospheric aerosols, and hence quantifying their hygroscopic properties is important for accurately parametrizing their climate and environmental impacts. Kuang et al. [33] reviewed the lab and field measurements of CCN activity of secondary organic aerosols (SOA) and highlighted the factors that affect the degree of organic aerosol hygroscopicity. These factors include RH, water solubility, degree of atmospheric aging that affects carbon chain length, functional groups, and oxygen-to-carbon (O:C) ratio, and surface tension changes due to the presence of surfactants and liquid–liquid phase separation (LLPS).

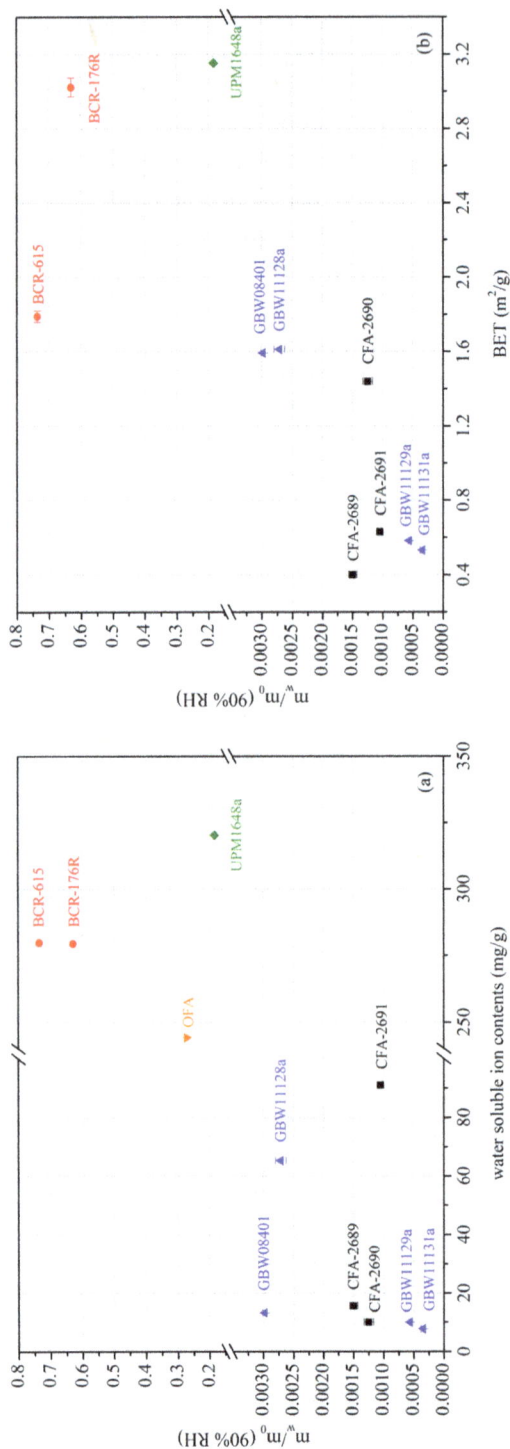

Figure 3.5: Dependence of mass ratios of adsorbed water to dry samples (m_w/m_0) at 90% RH on (a) water-soluble ion contents and (b) BET surface areas for the 11 unconventional mineral dust samples. The sample names are written next to the markers and they refer to: BCR-176 R and BCR-615 (European municipal waste fly ash), CFA-2689, CFA-2690 and CFA-2691 (American coal fly ash), GBW08401, GBW11128a, GBW11129a, and GBW11131a (Chinese coal fly ash), UPM1648a (National Institute of Standards and Technology (NIST) road dust sample), and OFA (oil fly ash from a heavy oil-fired boiler). Reprinted (adapted) with permission from reference [29], © American Chemical Society, 2020.

To highlight the contribution of the aforementioned factors, the hygroscopic properties of lab-generated semisolid SOA was measured as a function of RH and oxidation state, by quantifying the O:C ratio [34]. The SOA were generated in a flow reactor by homogenous nucleation and condensation following OH and/or O_3 oxidation of gas-phase precursors: isoprene (C_5H_8), α-pinene ($C_{10}H_{16}$), and longifolene ($C_{15}H_{24}$). The hygroscopicity parameter, κ, was measured in the sub- and super-saturation RH regions, using a hygroscopicity tandem differential mobility analyzer (HTDMA) and a cloud condensation nuclei counter (CCNc), yielding $κ_{HGF}$ and $κ_{CCN}$, respectively. The authors calculated the ratio of $κ_{HGF}$ to $κ_{CCN}$ between 0 and 1 to describe a general hygroscopic behavior, which is independent of absolute κ-values. κ-ratios below 0.5 indicate the dominance of water surface adsorption, whereas those above 0.5 indicate increasing solubility and higher water uptake. The phase of the SOA particles (solid, semisolid, liquid) was quantified using an aerosol bounce instrument that yields a particle bounced fraction (BF). Particles with BF>0 are solid or semisolid, and the particles with BF = 0 behave mechanically as liquids [34]. For comparison, similar experiments were conducted using solid and insoluble SiO_2 spheres, where water uptake occurs via surface adsorption. Figure 3.6 shows a schematic that summarizes their major findings: for sub-saturated conditions (RH<100%), the water uptake of SOA with relatively low O:C (<0.6) is an adsorption-dominated process. With increasing atmospheric aging due to oxidation, and increasing RH, dissolution increases, and the κ-ratio approaches 1, similar to supersaturation conditions. The authors also found that the organic precursor has an effect on the hygroscopic properties, even if the SOA O:C ratio is similar. For example, for O:C ratio of ~0.6, α-pinene SOA would have a κ-ratio of 0.82, whereas the κ-ratio for longifolene SOA is around 0.4.

Moreover, in addition to dissolution, type of volatile organic compound (VOC) precursor, and RH, Rastak et al. [35] found that LLPS can take place in some SOA originating from biogenic emissions with increasing RH. Figure 3.7 shows optical microscopy images for micrometer-scale isoprene- and α-pinene-SOA generated in the lab from photo oxidation and ozonolysis, respectively. The SOA products were collected on a hydrophobic glass slide for collecting images at 290 K in a temperature- and RH-controlled flow cell. The presence of multiple phases is visible in the optical images. One single phase was observed over the entire RH range for isoprene-SOA as shown in Figure 3.7A, whereas for α-pinene-SOA, a single organic-rich phase is observed below 95% (Figure 3.7B). Increasing RH above 95% results in phase separation, where a water-rich phase is observed. The authors found agreement between the microscopy images and the water uptake of considerably smaller 100 nm SOA particles generated in the lab from isoprene and α-pinene. It was noted that with atmospheric aging, the O:C ratio of monoterpene-SOA, its hygroscopic and CCN activity would resemble that of more oxidized isoprene-SOA. Improved thermodynamic model predictions of $κ_{HGF}$ were obtained when LLPS was included to explore the effect of surface tension reduction by organics, as well as the coupled gas-particle partitioning of semi-volatile organic molecules and water upon RH changes [35]. In the case of secondary marine aerosol consisting of

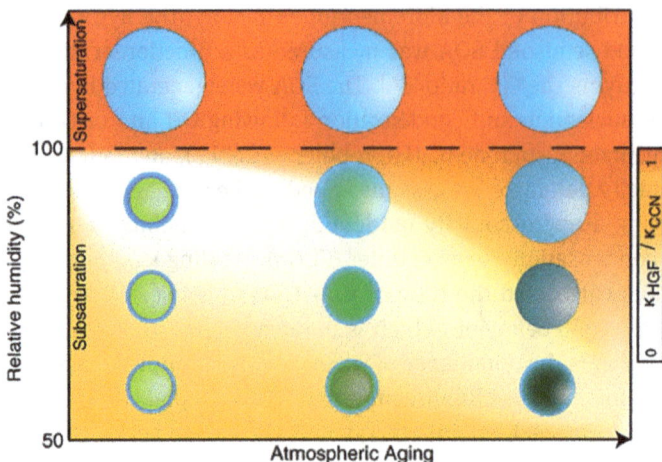

Figure 3.6: Schematics of the water-uptake processes of SOA particles in the atmosphere. Sub-saturated swelling (κ_{HGF}) can vary dramatically with minor differences in supersaturated droplet activation (κ_{CCN}); consequently particles can have very different direct and indirect climate effects. The background color scale indicates the ratio of these parameters at given sub-saturated and supersaturated conditions, whereas the darkness of the green color in particles denotes their atmospheric age. The contrast is largest when adsorption is the dominant water-uptake mechanism, even at high RH. This is the case for low O:C SOA particles on the left. With increasing oxidation, i.e., increasing atmospheric age of the particles, the solubility increases and the dissolution RH decreases, decreasing the discrepancy between the κ_{HGF} and κ_{CCN} values. Reproduced with permission from reference [34] under Create Commons License (CC BY-NC-ND 4.0) © The Authors, 2015.

sulfate, ammonium, and organic species, Mayer et al. [36] found that the measured CCN activity ($\kappa_{app} = 0.59 \pm 0.04$) of the resulting secondary marine aerosol matched the values observed in previous field studies, because these secondary particles correlate with phytoplankton biomass (i.e., chlorophyll-a concentrations), and primary sea spray aerosol does not. Importantly, the authors concluded that secondary marine aerosols play the dominant role in affecting marine cloud properties [36].

In addition, aromatic and aliphatic dicarboxylic acid compounds detected in field-collected SOA were shown to be reactive in the presence of transition metals such as iron. Iron-catalyzed aqueous phase reactions with catechol, guaiacol, fumaric, and muconic acids produced water-insoluble and light-absorbing secondary organics particles, namely polycatechol, polyguaiacol, Fe-polyfumarte, and Fe-polymuconate [37, 38]. Reaction conditions were chosen to mimic those taking place in adsorbed water under acidic conditions. Rahman and Al-Abadleh [39] studied the hygroscopic properties of these particles using DRFITS and quartz crystal microbalance (QCM) as a function of RH, at 298 K. A modified three-parameter Type II multilayer BET adsorption model described the water adsorption isotherm on the nonporous materials, polycatechol, polyguaiacol, and Fe-polymuconate, whereas, a Type V adsorption model (the

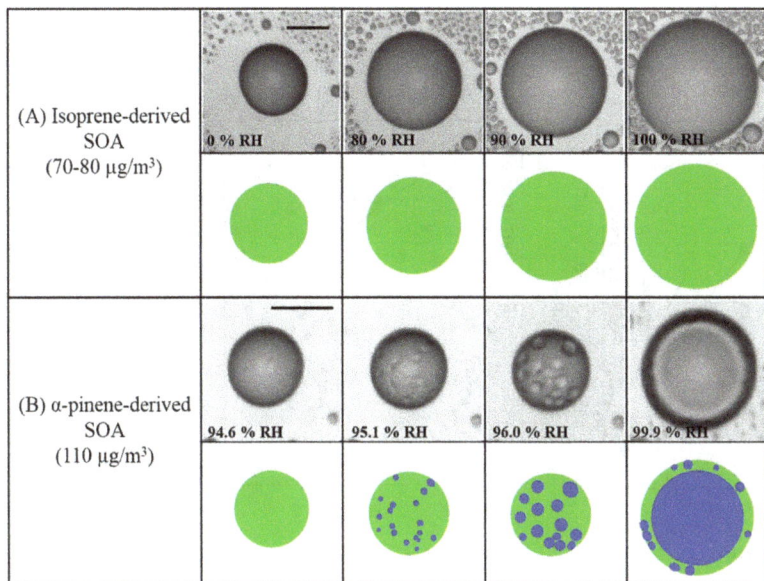

Figure 3.7: Optical images of micrometer-scale SOA particles with increasing relative humidity. (A) Isoprene-derived SOA for mass concentrations of 70–80 µg m^{-3} and (B) α-pinene-derived SOA for a mass concentration of 110 µg m^{-3}. Note that the light gray circles at the center of the particles are due to an optical effect caused by the hemispherical shape of the particles deposited on a substrate. Illustrations are shown below the images, for clarity. Green: organic-rich phase. Blue: water-rich phase. The scale bar represents 20 µm. Reproduced under Create Commons License from reference [35], © The Authors, 2017.

Langmuir-Sips model) that accounts for condensation in pores was used in the case of porous Fe-polyfumarate. Figure 3.8 shows the water adsorption isotherms on these materials with desorption data as well. The analysis of these data show that organometallic polymers are more hygroscopic than organic polymers and can retain more water with decreasing RH, due to structural differences.

3.1.2 Structure of surface water

There are molecular level structural differences between bulk and interfacial water on surfaces of atmospheric relevance that have consequences on the extent of water uptake, ice nucleation, ionic mobility, and chemical reactivity. These structural differences arise from changes to the hydrogen bonding network in the condensed phases of water that affect the fundamental vibrational modes of water [40]. The use of experimental and computational tools such as infrared spectroscopy, nonlinear sum frequency generation spectroscopy, X-ray based techniques, and molecular dynamics simulations provided invaluable information on the structure

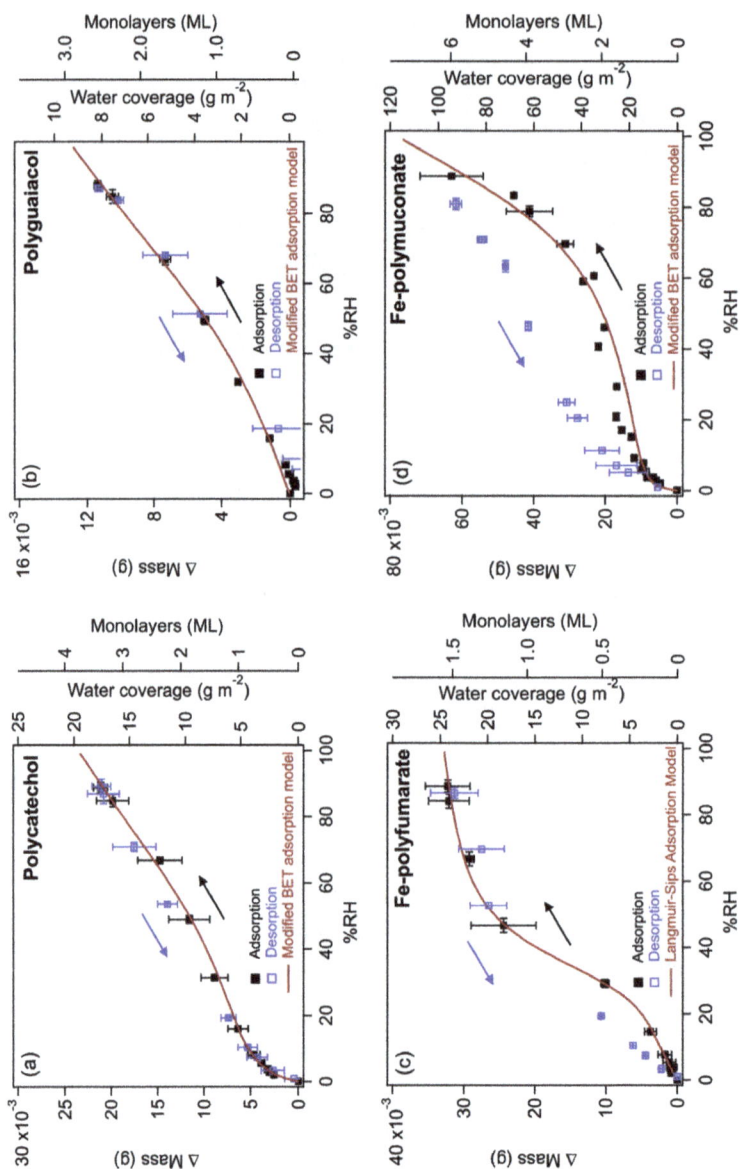

Figure 3.8: Water adsorption isotherms at 298 K for (a) polycatechol, (b) polyguaiacol, (c) Fe-polyfumarate, and (d) Fe-polymuconate. The left axis shows Δ mass (m), which is the change in mass due to adsorbed water, as measured using the QCM. The right axes were calculated by converting Δm to water mass (g) per surface area (m²) of the organic film deposited. The lines through the data represent least-squares best fits to the experimental data using the modified BET and the Langmuir-Sips adsorption models. The outermost right axis was obtained by dividing water coverage in g m⁻² by that at the monolayer coverage obtained from the best-fit parameters (see reference [39] for details). Reproduced under an ACS AuthorChoice License from reference [39], © The Authors, 2018.

of interfacial water. These differences were the subject of thematic special issues of the *J. Phys. Chem. C.* [41] and *Phys. Chem. Chem. Phys.* [42]. In the next few sections, major highlights from the above surface science tools and theoretical approaches are provided.

3.1.2.1 Hydrogen bonding network from infrared spectroscopy and sum frequency generation

Formation of hydrogen bonding in water clusters, liquid and solid phases affect the frequencies and oscillator strengths assigned to stretching, bending, and libera-tional modes of water [40, 43–45]. Also, heterogeneity (i.e., ordered vs. disordered) of the molecular environment of hydrogen bonded water can be assessed through the OH-stretching infrared band shape, which becomes broader with increasing dis-order [45]. The average hydrogen bond coordination number for water molecules in bulk water is ~3.5, which drops to 3.0 at the interface [46, 47].

Table 3.2 lists the values of the vibrational modes of water in its three phases, which aid in the interpretation of spectra collected for surface water. Using isotopi-cally labeled hydrogen and oxygen also aid in elucidating the structure of liquid water while analyzing the infrared spectra of H_2O, D_2O and HDO [48]. In general, for-mation of hydrogen bonds in the liquid and solid bulk phases leads to the formation of a broad OH-stretching (v_1, v_3) and H_2O-bending (δ) modes, and the appearance of a new combination mode. Hydrogen bonds are random and disordered and weaker in the liquid phase compared to ice, and that leads to shifting the band center from ~3400 to ~3200 cm^{-1} [49, 50]. For surface water, spectral features are often described as "liquid-like" and "ice-like", depending on the interface and the underlying substrate. A more accurate description refers to the strength of the hydrogen bonds: strong hydro-gen bonds give rise to broad spectral features lower than 3400 cm^{-1}, and weak hydro-gen bonds, between 3400–3500 cm^{-1}. In the case of no hydrogen bonds, the spectral features appear sharp between 3600–3750 cm^{-1} [51]. On a hydrophilic solid surface, such as hygroscopic salts and hydroxylated metal oxides and clays, water structure generally appears as a mixed ordered and disordered hydrogen-bonding network be-cause surface charges and, hence, surface field, can induce more polar ordering of interfacial water molecules that can extend to a few monolayers [52]. Water adsorbed on surfaces giving rise to "ice-like" peak structures such as in mica(001) were inter-preted to indicate that water molecules in direct contact with mica (i.e., first layer of water) are very tightly bound in a structured environment. Water molecules in this environment influence the ones in subsequent layers to reorganize and accommodate the crystalline structure of ice, because they are not as tightly bound. Hence, the pseudo hexagonal symmetry and lattice constants of mica(001) being close to ice do not directly influence mica's ability to act as a heterogeneous ice nucleating agent [45]. Figure 3.9 shows representative FTIR spectra of thin water films at other mineral surfaces that include metal (oxyhydr)oxides, clays, silicates, carbonates, and other natural specimens, which were collected at 38% RH at 25 °C [53]. The number of water "monolayers" at that RH varies, depending on the material and the particle size and range from near monolayer to multiple ones, per the isotherms measured in reference [53]. The authors [53] noted that micrometer-sized particles exhibit highly comparable distributions of an intense OH-stretching spectral features ~3400 cm^{-1} and a much less intense one ~3200 cm^{-1}, suggesting that mineral-bound water molecules form a smaller number and weaker hydrogen bonds than liquid water. Submicron particles, on the other hand, as in the case metal (oxyhydr)oxides, show more structured peaks and influence of water adsorption on surface hydroxyl groups.

Table 3.2: Summary of selected studies that utilized infrared spectroscopy for characterizing water in bulk phases, at interfaces, and surfaces of atmospheric relevance.

Bulk phase of water	Vibrational modes (cm^{-1})				Ref.
	Stretching		Combination $(\delta + v_R - v_T)$	Bending (δ)	
	Symmetric (v_1)	Asymmetric (v_3)			
H_2O (g)	3656	3755	–	1594	[43]
HDO (g)	3622 (OH) 2723 (OD)	3699 (OH) 2781 (OD)	–	1265	a
H_2O (l)	3261	3351	2134	1639	[58]
$H_2{}^{18}O$ (l)	3241	3337	2130	1632	[58]
HDO (l)	2502	3404	2900 (2δ)	1450	[59]
D_2O (l)	2407	2476	–	1206	[58]
H_2O (s)	3289	3411	2243	1641	[43]
Interface	**Interfacial water vibrational modes (cm^{-1})**				
Air/H_2O(l) or salt solutions (NaF, NaCl, NaBr, NaI)	3700 (dangling OH) 3645, 3755 (solvation, with salt solutions) 3400 (liquid-like) 3200 (ice-like)		–	–	[52, 60, 61]
H_2O(l)/OTS^b/fused silica interface	3680 (dangling OH) 3400 (liquid-like) 3200 (ice-like, more prominent)		–	–	[52, 56]

Table 3.2 (continued)

Bulk phase of water	Vibrational modes (cm^{-1})				Ref.
	Stretching		Combination $(\delta + v_R - v_T)$	Bending (δ)	
	Symmetric (v_1)	Asymmetric (v_3)			
$H_2O(l)$/hexane	3700 (dangling OH) 3400 (liquid-like) 3200 (ice-like)		–	–	[52]
$H_2O(l)$/CCl$_4$	3618, 3708 (dangling OH) 3500 (liquid-like) 3200 (ice-like)		–	–	[52]
$H_2O(l)$/α-quartz(0001) or alumina or CaF$_2$ or silica or silica/TiO$_2$ nanoparticles	3400 (liquid-like) 3200 (ice-like) Relative intensity is pH-dependent and point of zero charge		–	–	[52]
$H_2O(s)$/α-quartz(0001)	3150 (ice-like)		–	–	[52]
Inorganic solid Surface (single crystals and powders)	**Adsorbed water vibrational modes (cm^{-1})**				
α-Al$_2$O$_3$(0001)	3547, 3400, 3280 3700, 3404, 3211		–	–	[43] [62]
α-Al$_2$O$_3$(11$\bar{2}$0)	3740, 3300, 3150 (pH 2)		–	–	[63]
CaCO$_3$(10$\bar{1}$4)	3605, 3572, 3432		–	–	[14]
NaCl(001)	~3500 (sub ML) ~3400 (liquid-like @ 1 ML)		–	–	[44]
Muscovite mica (001) $(K^+[Al_2(Si_3Al)O_{10}(OH)_2]^-)$	~3350 (sub ML, cluster-like) ~3400 (liquid-like @ 1 ML)		–	–	[45]
MgO (001)	~3720 (-ve, OH$^-$ $_{surf}$) ~3400 (liquid-like)		–	–	[45]
BaF$_2$ (111)	~3250, 3450 (ice-like bilayer, < 1)ML ~3400 (liquid-like, > 1 ML)		–	–	[45]

Table 3.2 (continued)

Bulk phase of water	Vibrational modes (cm^{-1})				Ref.
	Stretching		Combination ($\delta + v_R - v_T$)	Bending (δ)	
	Symmetric (v_1)	Asymmetric (v_3)			
Powder metal (oxyhydr) oxides, clays, silicates, carbonates, natural specimens) Geothite (α-FeOOH), Lath Lepidocrocite (γ-FeOOH), Lath Lepidocrocite (γ-FeOOH), Rod Lepidocrocite (γ-FeOOH), Akaganeite (β-FeOOH), Ferrihydrite (Fe$_{8.2}$O$_{8.5}$(OH)$_{7.4}$ · 3 H$_2$O), Hematite (α-Fe$_2$O$_3$, 10 nm), Hematite (α-Fe$_2$O$_3$, 5 µm), Gibbsite (γ-Al(OH)$_3$), Kaolinite, Illite, Montmonrillonite, Quartz (α-SiO$_2$), Microcline (K (AlSi$_3$O$_8$)), Olivine ((Fe,Mg)$_2$SiO$_4$), CaCO$_3$, Volcanic ash, ATD	@ 38% RH, 25 °C ~3400 ~3200		–	–	[53]
α-Al$_2$O$_3$ powder	3570, 3420, 3200		2136	1642	[43]
γ-Al$_2$O$_3$ powder	3583, 3444, 3188		2128	1646	[43]
ATD	~3417 (liquid-like) ~3200 (< 1 ML, ice-like structures from coordination of water molecules to the cations)		2167	1640	[26]
Kaolinite	3417		–	1641	[12]
Illite	3631, 3398		–	1641	
Montmorillonite	3622, 3394		~2100	1633	
SiO$_2$	3744 (-ve), 3600, 3251		2139	1635	[9]
α-Al$_2$O$_3$	3740 (-ve), 3600, 3286		2153	1643	
TiO$_2$	3401		2129	1645	
γ-Fe$_2$O$_3$	3690, 3589 [c], 3113 [c]		2097	1640	
CaO	3335		2180	1640	
MgO	3333		2168	1645	
Coal fly ash	3673, 3400, 3240		2080–2160	1640	[31]
CaCO$_3$	3596, 3372, 3247		2100	1646	[64]

Table 3.2 (continued)

Bulk phase of water	Vibrational modes (cm^{-1})				Ref.
	Stretching		Combination $(\delta + v_R - v_T)$	Bending (δ)	
	Symmetric (v_1)	Asymmetric (v_3)			
Organic solid Surface (powder, films)	Adsorbed water vibrational modes (cm^{-1})				
Pollen	3593, 3205		2135	1616	[55]
Tannic acid[d]	3600, 3400, 3169		–	1645	[54]
Humic acid/goethite	3582, 3512, 3491		–	1641	[65]
Polycatechol Polyguaiacol	3580, 3020 3580, 3170		–	1628 1628	[39]
Fe-polyfumarate Fe-polymuconate	3641, 3564, 3020		–	1632 1628	[39]
C8 and C18 self-assembled monolayers	~3400 ~3180		–	–	[56, 57]

Notes: [a] Spectrum collected in the author's lab. [b] OTS = octadecyltrichlorosilane. [c] Assigned to FeO (OH) groups. [d] Water adsorption leads to the formation of organic hydrates.

In the case of hydrophobic surfaces, such as hexane and CCl_4, the spectra of water show features assigned to free and bonded OH groups, suggesting strong orientational effects [51]. These features were also apparent on tannic acid [54], polycatechol, polyguaiacol, Fe-polyfumarate, Fe-polymuconate [39], and pollen [55]. The weak and strong hydrogen-bonding networks of adsorbed water on these organic powdered materials suggest cluster formation, reflecting water–water and water–organic interactions. Also, water bonding with organic functional groups acting as hydrogen bond acceptors caused shifts in their vibrational modes [39] and led to partial dissolution due to the formation of hydrates, as observed in the case of tannic acid [54]. Water molecules in contact with organic self-assembled monolayers (SAMs) at low RH form clusters, which have fewer hydrogen bonds and gave rise to spectral components at 3200 cm^{-1}, whereas molecules in the interior of water clusters have 3 and 4 hydrogen bonds, similar to that of bulk water [56, 57].

3.1.2.2 Surface oxygen-hydroxyl binding energies from X-ray-based techniques

The application of X-ray based techniques, such as Near Ambient Pressure X-ray Photoelectron Spectroscopy (NAP-XPS), synchrotron-based X-ray Absorption Near

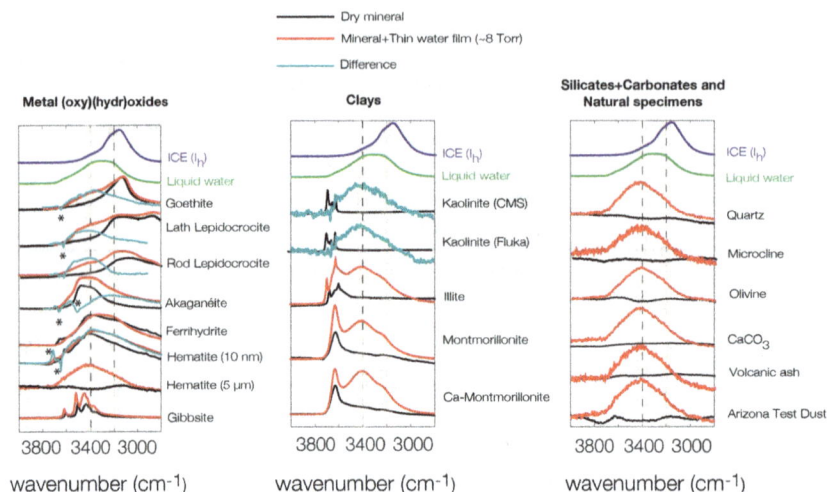

Figure 3.9: Representative FTIR spectra of thin water films at mineral surfaces at 25 °C. Reproduced with permission under Create Commons License (CC BY) from reference [53], © The Authors, 2016.

Edge Spectroscopy (XANES), and Extended X-Ray Absorption Fine Structure Spectroscopy (EXAFS) in atmospheric science provided invaluable insights of the structure of surface water and mechanisms of heterogenous reactions [66–70]. Data from X-ray-based experiments coupled with theoretical calculations for water in the bulk liquid and solid phases provided insights into the effect of hydrogen bonding on the electronic structure of water [71]. Water adsorption on surfaces of atmospheric relevance studied using NAP-XPS include α-Fe_2O_3(0001) [72], amorphous SiO_2, and TiO_2(110). In very general terms, for a given element, the binding energy of core electrons increases with increasing oxidation state of metals. In the case of anions, like O^{2-}, a decrease in the negative charge increases the binding energy of core electrons: "ion O^{2-}" in the oxides of transition and alkaline earth metals have binding energies between 527.7–530.6 eV, "ion OH^-" in the peroxides and hydroxides of transition and alkaline earth metals between 530.6–531.1 eV, "ion O^-" between 531.1–532 eV, and weakly adsorbed species between 531.1–533.5 eV [73]. For water adsorption on α-Fe_2O_3(0001) at 295 K [72], the O 1s binding energy spectra show distinct features assigned to lattice oxygen (~530 eV), hydroxyl groups (531.5 eV), molecular adsorbed water (533.3 eV), and gas phase water (535.5 eV), at relatively high vapor pressures. Data revealed that hydroxylation occurs at very low RH as a result of water-catalyzed dissociation at the topmost surface, followed by adsorption of molecular water with increasing RH.

For water adsorption on hydroxylated-SiO_2 [69], molecular adsorption of water occurs over a wide RH range, with rapid kinetics between 0 and 20%RH till 2 monolayer (ML) of water are formed. A slower increase in the rate of adsorption occurred between 20 and 80% RH, leading to the formation of 4–5 ML. Surface potential measurements increased in that RH range till a final value of +0.4 V was reached, suggesting that water bound to surface OH groups is randomly oriented as in bulk water with surface molecules that have dangling OH bonds pointing to the vacuum. The NAP-XPS data for the adsorption of water on $TiO_2(110)$ [69, 70] showed that the surface has a small number of bridge oxygen vacancy defects (V_{bridge}), where water adsorbs dissociatively in a fast step, leading to the formation of two hydroxyl groups according to the reaction: $H_2O + V_{bridge} + O_{bridge} = 2OH_{bridge}$. These groups act as the nucleation sites for subsequent molecular adsorption of water, via hydrogen bonding.

Using NEXAFS, the structure of liquid water and ice was studied by analyzing the O K-edge spectral features [74, 75]. In these experiments, the O1s electron is excited to the empty electronic states in the conduction band, and the recorded spectral features in the energy range 530–550 eV reflect the 2p character of the unoccupied valence orbitals of O in the water molecule [75]. Three regions are typically analyzed in the observed spectra: pre-edge (around 535 eV), main edge (537–538 eV), and the post edge (540–541 eV). In very general terms, energy values increase with increased number and strength of coordination to neighboring atoms/molecules. In their analysis of spectra collected for bulk ice, surface ice, and liquid water, Wernet et al. [74] showed that the intensity of the post-edge feature is dominant in the bulk ice spectrum, whereas for surface ice and liquid water, the main edge is dominant with a well-defined pre-edge, and a less intense post-edge. The intensities in the pre- and main edge were assigned to water molecules with one uncoordinated OH group, whereas the post-edge intensity was assigned to fully coordinated water molecules [74]. This observation suggests that, at room temperature, bulk liquid water has local coordination comparable to that at the ice surface, with OH groups involved in one strong hydrogen bond and none or weak hydrogen bonds. Increasing the temperature of pure water disturbed the hydrogen bonding network of water, where the intensity of the pre- and main edge increases and that in the post-edge decreases [74]. Adding alkali metal halides to liquid water resulted in similar observations in the oxygen K-edge X-ray absorption spectra, but the intensity changes in the main edge were found to arise from direct perturbation of the unoccupied molecular orbitals in water by the anions [76]. The combination of direct ion-water electrostatic interactions and geometric rearrangement of the hydrogen-bonding network explained changes in the pre- and post-edge intensities [77]. Using the same technique to study the surface of frozen NaCl solution, the intensities were found to be temperature-dependent and consists of ice, different phases of NaCl (NaCl, $NaCl \cdot 2H_2O$), and surface-adsorbed water [77].

3.1.2.3 Visualizing surface water structure using computational methods
The structure of water at interfaces and adsorbed on surfaces of different chemical composition was studied using density functional theory calculations and molecular

dynamic (MD) simulations. Recent advances in computational analysis of the first monolayer of adsorbed water and the effect of that layer on subsequent layers were recently reviewed [78, 79]. The metal oxides investigated include TiO_2, SiO_2, and MgO of different surface planes. The major findings were as follows: (1) water interacts preferentially with unsaturated metal sites via its O atom; (2) part of the water molecules partially dissociate after adsorption on TiO_2 and MgO surfaces, depending on the surface coverage and underlying surface plane; and (3) most water molecules adsorb molecularly on the hydroxylated SiO_2 surface, with the location of O atoms and orientation of H atoms forming well-defined hydrogen bonding networks. The presence of mono- and di-valent cations with chloride as the counter anion were found to perturb the hydrogen bonding network of water at the surface of SiO_2 (101) [80] under simulated acidic and neutral conditions relevant to atmospheric conditions. This perturbation was evident from the apparent structure in the probability profile of O and H in water molecules in the presence of ions compared to pure water, and in the reorientation of water molecules away from the surface to minimize hydrogen bonds with surface groups. Simulated water adsorption on CaF_2(111) and CaF_2(100) showed that the modestly hydrophobic (111) surface (contact angle ~20°) results in a gap between surface fluoride ions and interfacial water molecules. The (100) surface, on the other hand, is hydrophilic and simulated water molecules fully hydrate the surface.

MD simulations were also conducted on goethite and hematite surface planes. For example, Song and Boily [81] calculated detailed information on the water density profile on hydroxylated goethite α-FeOOH (110) plane as a function of distance from the hydroxylated surface. Figure 3.10 shows the major findings of these simulations. From the data in Figure 3.10a, two predominant adsorbed water layers coexisted, with the first layer about 0.08–0.11 nm from the top –OH groups. The water molecules in these two layers were oriented 60–70° from the normal plane, with protons oriented toward the surface as a result of hydrogen bonds (HB) with –OH, μ–OH and $\mu_{3,\mathrm{II}}$–OH groups. Additional analysis of HBs is provided in Figures 3.10c and d, showing the HB population between water and surface O and water–water interactions (i.e., fraction for a given H_2O surface coverage). This analysis shows that surface O atoms are predominantly H-bond acceptors, and water–water interaction is dominated by 1–2 HB, leading to the conclusion that liquid water-like films form on goethite. In the case of water thin films on hematite, Boily et al. [82] reported MD simulations on multifaceted α-Fe_2O_3 nanoparticle surfaces (100), (001), (110), (012), and (014). The results showed that the stabilization of thin water films is enabled by the majority of hydrogen bond-donating –OH groups, and that water–water HB populations are greatest on the (001) face, and decrease in importance in the order (001) > (012) ≈ (110) >(014) ≫ (100). Also, the calculations showed that sites at particle edges interconnect thin water films grown along adjacent crystallographic faces.

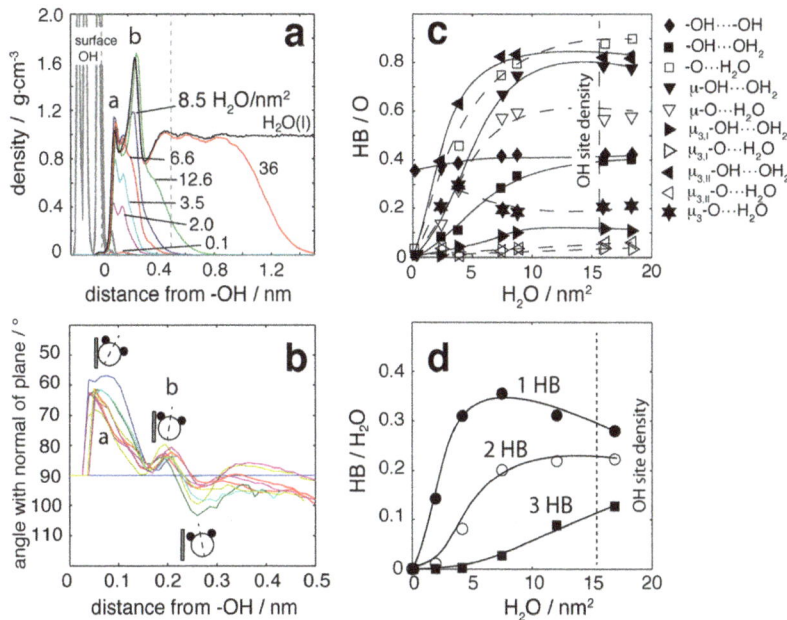

Figure 3.10: Water density profiles (a) and average water orientations (b) are shown as a function of the distance from the top −OH groups of the (110) plane. Schematic representations of the angles formed with the normal of the surface are included in (b). MD-derived H-bonding populations for water-surface (hydr)oxo (c) and water–water (d) interactions. The total crystallographic (hydr)oxide site density is shown by the vertical dashed line in (c) and (d). H-bond was defined on a geometric basis as hydrogen-acceptor distances below 0.3 nm and hydrogen-donor–acceptor angles below 30°. Reprinted (adapted) with permission from reference [81], © American Chemical Society, 2013.

The interaction of water molecules with organic substrates was also investigated using MD simulations [57, 83]. Focusing on two systems relevant to the formation of carboxylic acids from the oxidation of alkanes, Moussa et al. [57] examined water-surface and water–water hydrogen bonding on -CH₃ and –COOH-terminated SAMs in the sub monolayer region. Figure 3.11A shows the results of these simulations on the -CH₃-terminated SAM (a model for a hydrophobic surface), where water clusters formed and coalesced into a larger droplet. The probability of having water molecules with one and two hydrogen bonds, P(1 +2), decreased as the probability of forming three and four hydrogen bonds increased, P(3 +4), with increasing number of water molecules. This result suggests the dominance of water–water interactions inside the cluster relative to water-surface interactions. In the case of –COOH-terminated SAMs (a model for a hydrophilic surface), the simulations in Figure 3.11B show that water clusters formed and are distributed across the surface, with increasing water molecules. The P(1 +2) for water-surface hydrogen bonding is highest at lower water coverages, and P(3 +4) for water–water interactions increase with increasing the number of water molecules.

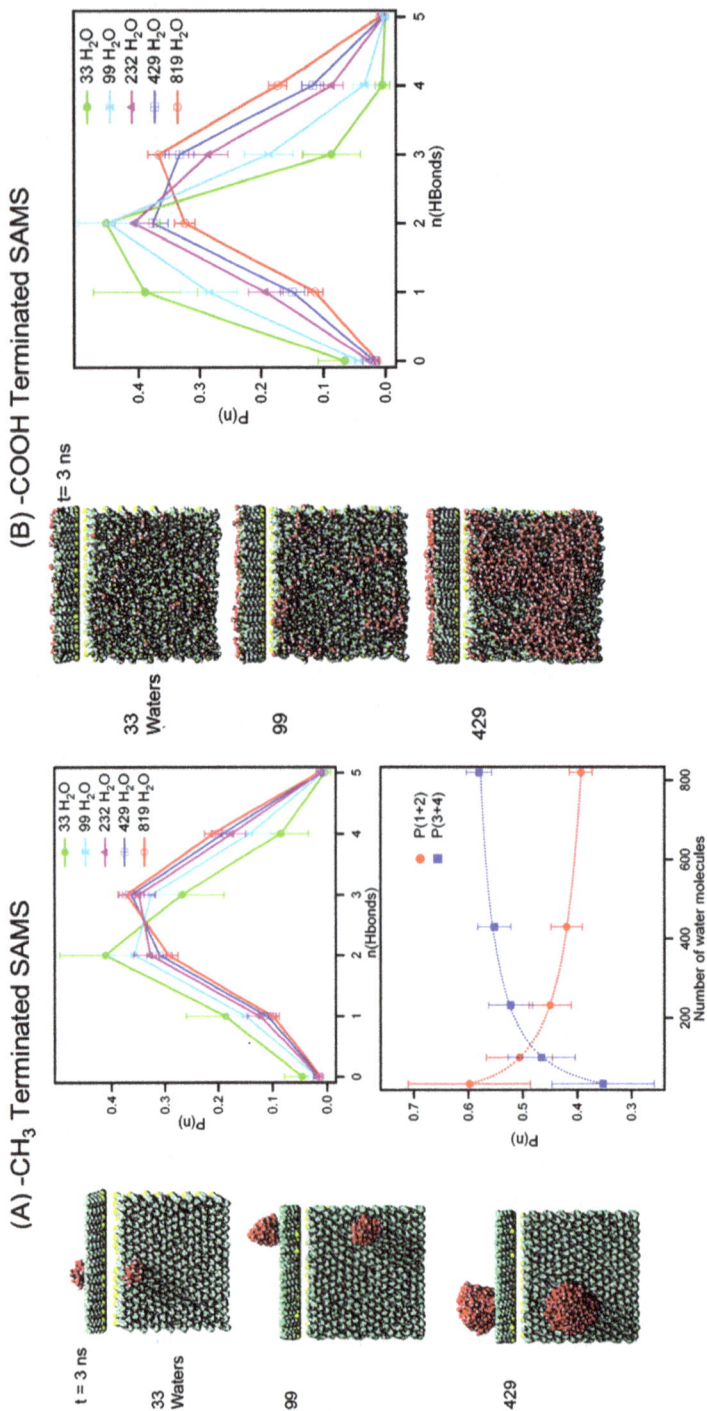

Figure 3.11: MD simulations of water on -CH$_3$ (left) and -COOH (right)-terminated C8 SAM showing side- and top-view snapshots for different numbers of water molecules at equilibrium. Also shown are probability distribution, P(n), of a water molecule forming a particular number of hydrogen bonds, n(Hbonds), which represents the total probability of hydrogen bonding to the substrate and to other water molecules. The left panel also shows P(n) as a function of the number of water molecules. Reprinted (adapted) with permission from reference [57], © American Chemical Society, 2009.

3.1.3 Gas phase water uptake on soluble aerosol components

Water uptake by pure salts, highly soluble organics, and mixtures of soluble salts and organics proceeds via different mechanisms than on surfaces of insoluble materials [84–86]. Water adsorption still occurs on dry particles at low RH, and phase transitions are observed at room temperature as a function of increasing and decreasing RH. Deliquescence (D) is a thermodynamically favorable solid- to liquid-phase transition that occurs at a specific RH (DRH). DRH can be predicted from the effect of solutes in saturated solutions on the vapor pressure of water at a given temperature [84, 85, 87, 88]. With decreasing RH, efflorescence (E) is observed, which is a kinetically controlled liquid- to solid-phase transition that occurs due to the decrease in the liquid water content, creating supersaturated solutions, until a nucleation site is formed for "precipitation" to take place [88]. Hence, a hysteresis effect is observed, where ERH ≠ DRH. As a physical process, efflorescence is rationalized by classical theories of homogenous and heterogeneous nucleation [88, 89]. The presence of insoluble impurities in deliquesced particles speeds up the efflorescence kinetics, leading to smaller gaps between ERH and DRH. As noted in the review by Krieger et al. [86], when organic substances are exposed to cycles in RH, their response can be divided to two classes: one class of substances shows deliquescence and efflorescence, and another class shows a continuous water uptake and release. In general, extent of water solubility determines if an ERH exists. For example, succinic acid has low aqueous solubility and tends to effloresce in single particle experiments; malonic acid, which is highly water soluble, tends to remain in a supersaturated metastable state, even at low RH [86]. The latter case could result in the formation of highly viscous particles at low RH that can even turn glassy. More details on aerosol viscosity are provided in the following sections. A list of the DRH and ERH for single, binary, and ternary inorganic salt systems is provided in reference [84] at 298 K. For inorganic/organic systems relevant to atmospheric particles, references [85, 86] provide values of DRH at 298 K. Table 3.3 lists these values for selected systems, where a range of values are shown for mixtures to indicate that variations were observed in DRH and ERH because of different molar/mass ratios and techniques used.

Depression in DRH and ERH were observed experimentally for the mixture of inorganic salt, salt/clay, salt/organic systems relative to pure salt systems. The extent of the depression depends on the relative molar mass ratio of the non-salt components, and for ERH, on the viscosity of the supersaturated system. For mixed organic/inorganic mixtures, a depression in DRH is observed when the addition of the organic substances increases the solubility of the salt per mass of water [86]. If adding the organic substance reduces the solubility of the salt, the DRH remains unchanged, but LLPS might be observed [86]. For example, Bertram et al. [88] showed that secondary organic material (SOM) produced from isoprene oxidation with O:C elemental ratio range from 0.68–0.74 reduced DRH and ERH of ammonium sulfate (AS) to a larger extent than organics produced from the ozonolysis of

Table 3.3: List of deliquescence and efflorescence relative humidity (%) for selected systems.

Inorganic salt systems [84, 85]			Soluble Organic systems [85–87]		
Composition	%DRH	%ERH	Composition	%DRH	%ERH
KCl	84–91	53–59	Citric acid (anhydr) ($\cdot H_2O$)	74–75 78	
KBr	81	52	Fumaric acid	>95	
K_2SO_4			Malic acid	59	
$Mg(NO_3)_2$ (ref. [91])	53±2	37–48	Potassium citrate monohydrate	59	
Na_2SO_4	84–93	55–57	Sodium benzoate	85	
NaBr (anhyd) ($\cdot 2H_2O$)	45 58	22	Sodium citrate dihydrate	83	
NaCl	75–77 69 (ref. [90]) 72–77 (ref. [86])	43–50 39–54 (ref. [86])	Sodium diacetate	70	
NaCl/KCl	73.8	38 for $x_{NaCl} > 0.4$ 64 for $x_{NaCl} < 0.4$	Sucrose	85	
NaCl/ montmorillonite	67 (ref. [90])		Xylitol	77–79	
$NaHSO_3$	80		Succinic acid ($C_4H_6O_4$)	98.9	52–59
$NaNO_3$	70–74.5	40 or not observed	Glutaric acid ($C_5H_8O_4$)	83.5–85	29–33
NH_4NO_3 (AN)	60 62 (ref. [86])	0–30 (ref. [86])	Adipic acid ($C_6H_{10}O_4$)	>99	>85
NH_4NO_3/ $(NH_4)_2SO_4$	77	18	Succinic acid/AS	79–80	30–48
$2NH_4NO_3 \cdot$ $(NH_4)_2SO_4$	56.4		Glutaric/AS	77–81	30–59
NH_4Cl	77	45	Adipic/AS	79–83	31–40

Table 3.3 (continued)

Inorganic salt systems [84, 85]			Soluble Organic systems [85–87]		
Composition	%DRH	%ERH	Composition	%DRH	%ERH
$(NH_4)_2SO_4$ (AS)	79–81 77 (ref. [90]) 80–81 (ref. [86])	33–48 34–48 (ref. [86])	Succinic acid/NaCl	73	53
$(NH_4)_2SO_4$/ montmorillonite	74 (ref. [90])		Glutaric/NaCl	71–75	30–49
NH_4HSO_4	39–41	not observed	Citric acid/NaCl	62	38–42
$(NH_4)_3H(SO_4)_2$	69	35	Succinic acid/AN	60–62	43–45
$(NH_4)_2SO_4$/ $(NH_4)_3H(SO_4)_2$	69		Levoglucosan/AN	not observed	not observed

α-pinene and β-caryophyllene, over the organic–sulfate mass ratio range 0.1–10 [88]. Within the uncertainty of the values, DRH of AS remained unchanged in mixtures with SOM from α-pinene and β-caryophyllene, whereas it shifted from 80 to 70% with increasing organic–sulfate mass ratio for SOM from isoprene oxidation. Similarly, values of ERH remained near 30% for the former system, and shifted from 25 to 20% for the latter. As an example for the salt–clay system, Atwood and Greenslade [90] reported that the DRH for internal mixtures of $(NH_4)_2SO_4$ and NaCl with montmorillonite shifted from 77% to 74%, and from 69% to 67%, respectively, due to the insoluble clay. These results were explained by changes in Gibbs free energy of the mixtures due to perturbations in ion–molecule interactions and lattice structure that results from the presence of montmorillonite.

3.2 Ice nucleation

Understanding the role of atmospheric aerosols in cloud and ice nucleation remains at the core of minimizing uncertainties associated with the indirect effect of aerosols on the climate and the hydrological cycle. Figure 3.12 links the role of aerosols in acting as CCN and IN to other microphysical processes that take place in a deep convective cloud. Selected recent reviews that synthesize the literature on the influence of aerosol chemical composition and atmospheric processing on CCN and IN activity can be found in these references [3, 92–94]. These reviews highlight the role of ice nucleating particles (INP) in heterogeneous freezing, which takes place at much higher (i.e., warmer) temperatures than that required for homogeneous nucleation

(< −38 °C or 235 K), and lower saturation with respect to ice (<1.5), as illustrated in Figure 3.13 [93]. Table 3.4 lists the brief definition of the different heterogeneous freezing processes per Kanji et al. [92].

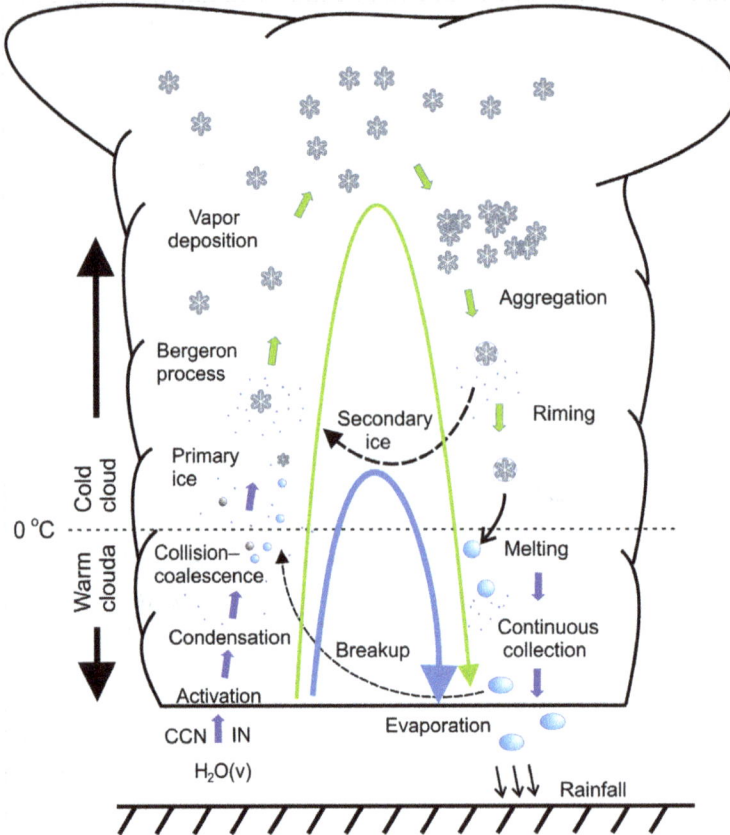

Figure 3.12: Summary of microphysical processes operating inside a deep convective cloud. Reproduced with permission from reference [95], © Elsevier, 2015.

Following decades of research, aerosols that make effective INP have been grouped into three categories: (a) insoluble/solid inorganic particles, (b) soluble particles, (c) organics and glassy particles, and (d) nanoscale biological fragments [92]. The chief examples of (a) are mineral dust, fly and volcanic ash, K-feldspars, metal and metal oxides, and soil dust, where experiments showed that factors affecting results include surface area, surface functional groups, surface defects, and lattice match of surface planes with that of ice. A few highlights from selected recent studies include the following: DeMott et al. [97] developed an empirical parameterization for the immersion freezing activity of natural mineral dust particles from both laboratory studies

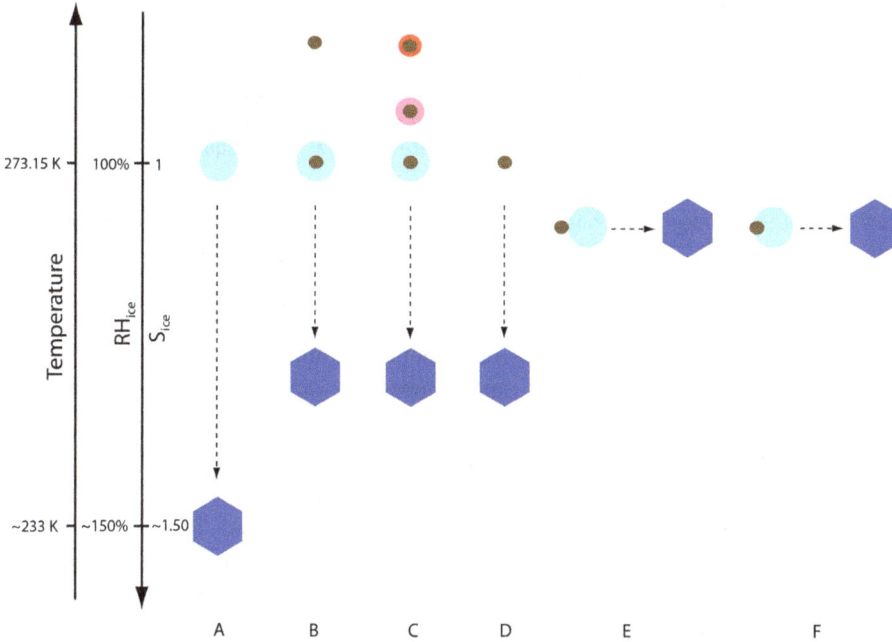

Figure 3.13: Schematic overview of different ice nucleation pathways, example given as a function of temperature, relative humidity, and supersaturation with respect to ice (RH_{ice} and S_{ice}, respectively). (A) Homogeneous ice nucleation, (B) immersion freezing, (C) deliquescence and water uptake followed by immersion freezing, (D) deposition ice nucleation, (E) contact ice nucleation, (F) inside-out freezing. Symbol forms and sizes not to scale. Reprinted (adapted) with permission from reference [93], © American Chemical Society, 2018.

Table 3.4: Brief definition of the heterogeneous freezing processes per Kanji et al. [92] and Knopf et al. [93].

Term	Definition
Immersion freezing	Ice nucleation initiated by an INP immersed in an aqueous solution or water droplet via activation of CCN during cloud formation. This process is suggested to be the most important for mixed phase clouds.
Deposition ice nucleation	Ice forms on an INP from the supersaturated gas phase (i.e., $RH_{ice} >$ 100%). Liquid water is presumed to be absent.
Condensation freezing	Freezing being initiated concurrently with the initial formation of liquid on CCN at super cooled temperatures.
Contact ice nucleation	Ice nucleation occurs when an INP collides or comes into contact with a super cooled water or aqueous solution droplet.

Table 3.4 (continued)

Term	Definition
Inside-out freezing	Freezing occurs when an INP is in contact with the liquid–air interface, establishing three interfaces (i.e., gas, solid, liquid), in super cooled conditions.
Pore condensation and freezing [96]	Ice formation is initiated at cirrus temperatures via the liquid phase in a two-step process involving condensation and freezing of super cooled water inside pores.

and atmospheric measurements. The dependence of immersion-mode ice nucleating ability of K-feldspars (containing 85% microcline ($KAlSi_3O_8$) and 15% albite ($NaAlSi_3O_8$)) on the surface composition was investigated by Yun et al. [98] to account for the effect of water-soluble inorganic solutes at low concentration. Their study found that only K^+ (aq) had a positive (i.e., warming effect) on the ice nucleating ability of K-feldspar, while the other alkali cations, namely Li^+, had no effect, and Na^+, Rb^+, and Cs^+ had a negative effect. This trend correlated with the K/Al ratio at the surface of K-rich feldspar. Whale et al. [99] investigated the immersion mode enhancement and suppression of ice nucleation of feldspars, humic acid, quartz, amorphous silica gel, and ATD with other alkali halide and ammonium salts. Chatre et al. [100] showed that increasing the mole fraction of citric acid when ATD is present raised the ice crystallization temperature compared to ATD only. In simulated dust aging experiments, Al-Abadleh and coworkers [101] showed that if catechol is added, ATD catalyzed the formation of polycatechol, which did not significantly impact ice nucleation or block ice nucleation sites on ATD in droplet freezing experiments. However, increasing pH reduced the ice nucleation ability of ATD. The effect of porosity of coal fly ash aerosol particles on their IN activity at cirrus temperatures was demonstrated by Umo et al. [96] to highlight the dominance of pore condensation and freezing processes in these systems.

In the case of soluble particles as INP, this class contains hygroscopic salts such as ammonium sulfate and sodium chloride, which can crystallize by efflorescence. The ice nucleating ability of these salts has been extensively studied and reviewed by Kanji et al. [92]. Laboratory studies showed that supermicron and submicron salt particles induce heterogeneous ice nucleation via contact freezing. Pure soluble organics such as citric acid were found to promote ice nucleation near homogeneous freezing temperatures, T < - 40 °C, when amorphous glass-like species form [92]. Studies to date have concluded that salt mixtures with organics affect the ice nucleating abilities of these systems in an unclear way because of variations in chemical composition and functional groups, phase, and morphology [102] compared to single component salt systems. In their modeling study, Yun and Penner [103] found that marine organic aerosols contribute to more ice formation than dust or black carbon/organic matter in mixed-phase clouds, and highlighted that their inclusion as natural heterogeneous ice nuclei reduced the magnitude of the total top-of-atmosphere anthropogenic aerosol forcing by 0.3 W m^{-2}. The review by Knopf et al. [93] examined, in detail, the current state of knowledge on the ice nucleating ability of organic matter in atmospheric aerosols, with additional details in reference [92] and highlighted the need for molecular level investigations to better understand the underlying mechanisms affecting organics ice nucleating abilities, particularly the structure of ice and nature of hydrogen bonding on different organic surfaces, as done for mineral clays and metal oxides [104].

3.3 Aerosol viscosity, volatility, mixing states, and morphologies

The grouping of aerosol viscosity (as a measure of phase state), volatility, mixing states, and morphologies is appropriate to emphasize the interdependency of these properties and that understanding one of them leads to better understanding the other ones. For SOA, which accounts for a large fraction of submicron particles in the atmosphere, their phase state (liquid, semisolid, solid) varies depending on a number of factors as detailed below. On a global scale, Shiraiwa et al. [105] quantified the global distribution of particle phase state, using the ratio of the glass transition temperature of (T_g) of SOA to modelled annual average of ambient temperature (T) in the year 2005–2009 (T_g/T), as an indicator of the particle phase state: $T_g/T \geq 1$, solid; $0.8 < T_g/T < 1$, semisolid; $T_g/T \leq 0.8$, liquid, at the earth's surface, 850 hPa (mbar), and 500 hPa. The value of T_g quantifies the phase transition between amorphous solid and semisolid states [106]. The results of this study highlighted that the particle phase state strongly depends on RH and ambient temperature, which determine the spatiotemporal distributions of SOA phase state worldwide. Wong DeRieux et al. [106] developed a method for the calculation of T_g and estimation of viscosity of individual SOA compounds (from α-pinene, isoprene, and toluene) and biomass burning particles with molar mass up to 1100 g mol^{-1}, using the number of carbon, oxygen, and hydrogen atoms measured experimentally. Within the uncertainties of this method, the two extremes in predicted RH-dependent viscosity values are for isoprene SOA (lower end) and α-pinene and toluene SOA (higher end), with biomass burning particles in between. For example, at 40% RH, the viscosity ranges for the two extremes are ~10–10^4 Pa s and 10^4–10^{10} Pa s, respectively, and the intermediate values are 10^3–10^6 Pa s. Comparison of predicted viscosities with measured values highlighted the effect of ionization techniques used in mass spectrometric instruments for elemental composition as one of the sources of the large uncertainties [106].

To investigate whether diffusion in "glassy" SOA is fast enough for phase partitioning to equilibrate, Ye et al. [107] conducted SOA mixing experiments to quantify the time scale for uptake of semi volatile constituents of toluene and α-pinene oxidation SOA as a function of RH. The results showed that toluene SOA contains about 30% semi volatile material ($C^* \geq 10$ μg·m^{-3}), whereas the remaining 70% has a volatility ($C^* \leq 1$ μg·m^{-3}), and that toluene oxidation SOA resists uptake of semi volatile vapors at very low RH. Above 40% RH, diffusion limitations vanish due to increased plasticity (i.e., decreased viscosity). On the other hand, α-pinene oxidation SOA took up substantial fraction of vapors from toluene oxidation SOA under dry conditions, and the extent of mixing at RH>40% was fast, similar to the toluene SOA case, i.e., mass transfer is not diffusion limited.

The quantification of mixing state of an aerosol population based on entropy was developed by Reimer and West [108]. Table 3.5 lists the definitions of the terms used in the calculation of the mixing state index, χ. The analysis done by Bondy et al. [109] shows the applicability of this index to quantifying the chemical mixing state of field aerosols. Figure 3.14 shows mixing state diagram for bulk diversity and average particle-specific diversity and mixing state indices for sub- and supermicron particles during the SOA, dust, and SSA periods of sample collections for the Southern Oxidant and Aerosol Study (SOAS) in Centreville, Alabama (a rural, forested location). The individual particles were analyzed for their elemental composition, size, and mass, so that elemental diversity can be calculated and correlated with size. The elemental diversity indicates each particle class: SOA/sulfate, biomass burning particles, fly ash, dust, nascent sea spray aerosols (SSA), and biological particles. The mixing state aging parameters that correlate mass values of elements characteristic to each particle class and to the mass of secondary species were also calculated. The conclusions made in Figure 3.14 indicate that the aerosol population analyzed was more externally mixed than internally mixed and that size and dominant particle class are among the factors that affect the degree of mixing state in atmospheric aerosols. Mixing state and morphology evolve together with atmospheric aging processes, as illustrated in Figure 3.15 by Xu et al. [110] during the movement of an Asian dust storm, which highlight the complexity in the quantification of these physical properties. Stevens and Dastoor [111] reviewed the state of atmospheric models in representing aerosol mixing state, focusing on the effects of simplified assumptions of aerosol mixing state on CCN concentrations, wet deposition, and aerosol absorption.

Figure 2.2B shows a schematic of major atmospheric aerosol mixing structures per You et al. [112], Li et al. [113], and Song et al. [114]. Figure 3.7 shows an example of the extent of LLPS in SOA from different precursors as a function of increasing RH from reference [35]. Similar studies were conducted on expanded lists of α-pinene and β-caryphyllene SOA particles and mixtures of commercially available organic compounds, with O:C elemental ratio in the range 0.13–1 [115]. The trend observed was that LLPS almost always occurred when the O:C was less than 0.44 and did not occur when O:C was greater than 0.44. The mechanisms of phase separation were described to occur by spinodal decomposition (as in Figure 3.7B), and nucleation and growth. Song et al. [114] studied similar systems in the presence of ammonium sulfate (AS). These studies found that AS in mixed organic/AS/H_2O particles deliquesced between 70 and 84% RH, and effloresced below 58% RH, or remained in a one-liquid-phase state, depending on the O:C ratio of the organics. In droplets with LLPS, AS always effloresced between 30 and 50% RH, which clearly suggested that the presence

Table 3.5: Definitions of aerosol mixing entropies, particle diversities, and mixing state index. In these definitions, we take $0\ln 0 = 0$ and $0^0 = 1$.

Quantity	Name	Units	Range	Meaning
$H_i = \sum_{a=1}^{A} -p_i^a p_i^a$	mixing entropy of particle i	–	0 to $\ln A$	Shannon entropy of species distribution within particle i
$H_\alpha = \sum_{i=1}^{N} p_i H_i$	average particle mixing entropy	–	0 to $\ln A$	Average Shannon entropy per particle
$H_\gamma = \sum_{a=1}^{A} -p^a p^a$	population bulk mixing entropy	–	0 to $\ln A$	Shannon entropy of species distribution within population
$D_i = e^{H_i} = \prod_{a=1}^{A} (p_i^a)^{-p_i^a}$	particle diversity of particle i	effective species	1 to A	Effective number of species in particle i
$D_\alpha = e^{H_\alpha} = \prod_{i=1}^{N} (D_i)^{p_i}$	average particle (alpha) species diversity	effective species	1 to A	Average effective number of species in each particle
$D_\gamma = e^{H_\gamma} = \prod_{a=1}^{A} (p^a)^{-p^a}$	bulk population (gamma) species diversity	effective species	1 to A	Effective number of species in the population
$D_\beta = \dfrac{D_\gamma}{D_\alpha}$	inter-particle (beta) diversity	–	1 to A	Amount of population species diversity due to inter-particle diversity
$\chi = \dfrac{D_\alpha - 1}{D_\gamma - 1}$	mixing state index	–	0 to 100%	Degree to which population is externally mixed ($\chi = 0$) versus internally mixed ($\chi = 100\%$)

of LLPS facilitates AS efflorescence. If the organics are dicarboxylic acids, then pH would also play a role in the ERH and separation RH (SRH). Freedman and coworkers [116, 117] demonstrated the pH dependency of salt phase transitions RH and LLPS in systems containing AS. In one study, 3-methylglutaric acid was used as a diacid in a system containing AS and NaOH [116]. Starting with fully mixed and deliquesced droplets followed by decreasing RH, the authors observed that increasing pH of the droplet from 3.7 to 6.5 slightly increased DRH (from 74 to 76%), reduced the SRH (from ~78 to 64%), and increased ERH (from ~40 to 52%). The results were explained as follows: increasing the pH increases the concentration of fully deprotonated organic species, which increases their solubility and prevents salting out of the organics (hence, separation occurs at lower RH). The authors hypothesized that upon separation at the higher pH, the organics form a seed crystal on which AS effloresces. Varying the ammonium to sulfate ratio (ASR) through the addition of sulfuric acid was also found to cause shifts in SRH [117]. The addition of sulfuric acid reduced the pH and the

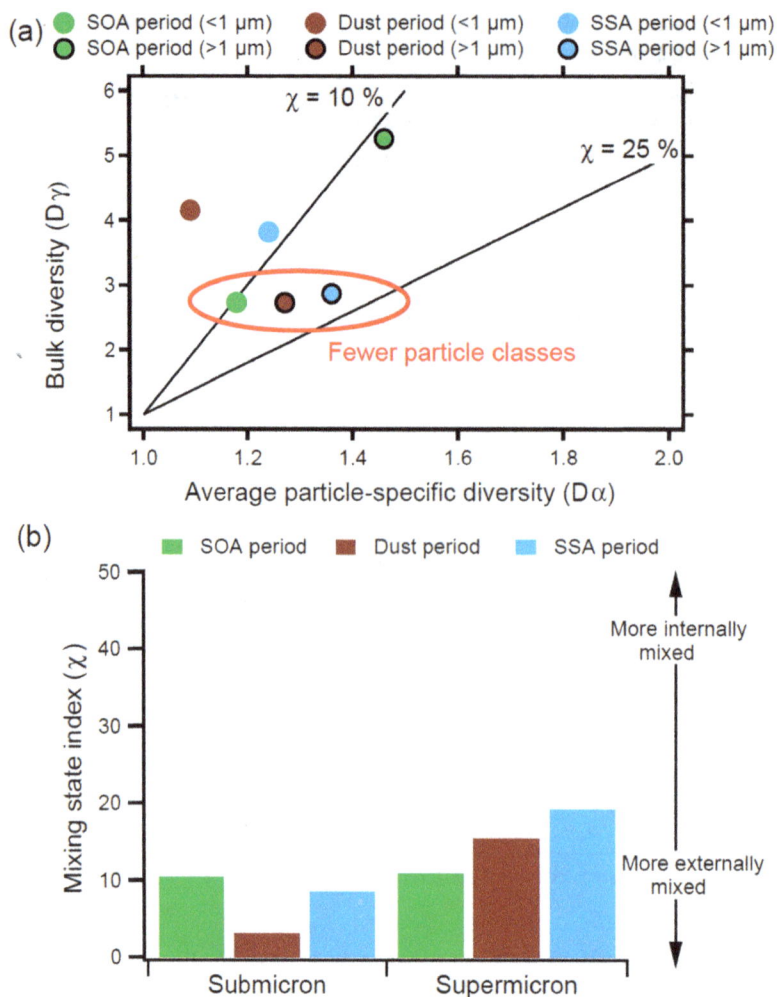

Figure 3.14: (a) Mixing state diagram showing the bulk diversity and average particle-specific diversity and (b) mixing state indices for sub- and supermicron particles during the SOA, dust, and SSA periods. For submicron particles, contributions by different sources impact mixing state. Reproduced under Create Commons License 4.0 (CC-BY) from reference [109], © The Author(s), 2018.

ASR to below 1, which resulted in lower SRH to no LLPS (as observed using 2,2-bis (hydroxymethyl)butaric acid and 1,2,6-hexanetriol).

While aerosol viscosity was not measured in references [35, 115] because the organics were liquid at room temperature, an example of a study that did that is the one by Hosny et al. [118], where the authors quantified microviscosity of SOA particles from the ozonolysis of myrcene (a common monoterpene from the biosphere), α-pinene, and oleic acid (a proxy for atmospheric alkene). Figure 3.16 shows data from a fluorescence-based

(A)

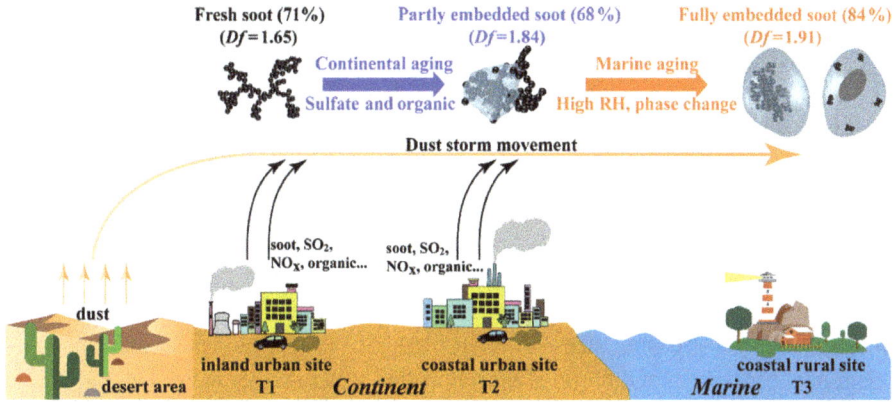

(B)

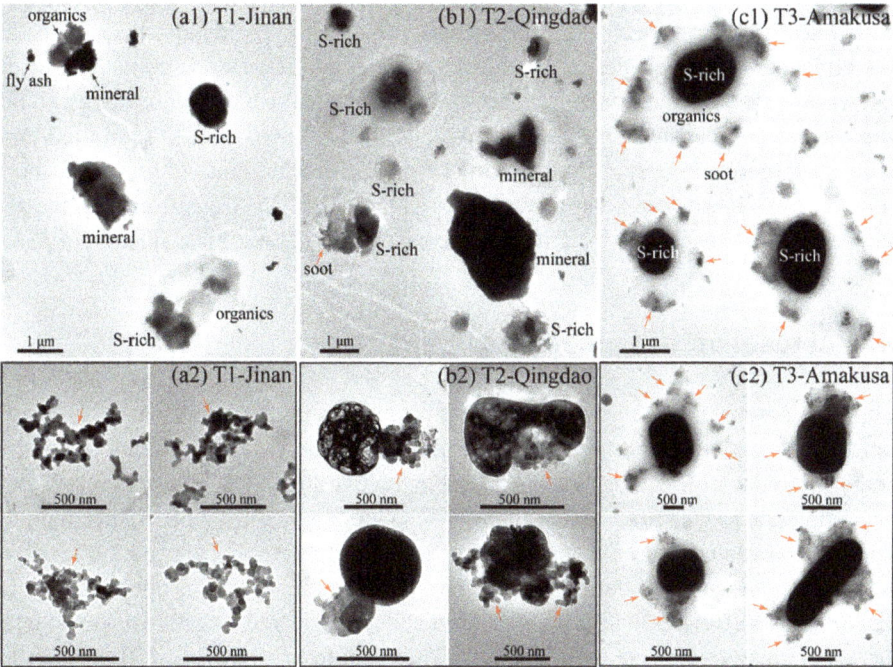

Figure 3.15: (A) Schematic diagram showing the evolution of morphology and mixing state of soot particles along with the movement of an Asian dust storm. (B) (a1–c1) Low-magnification TEM images at T1, T2, and T3; (a2–c2) TEM images of soot-dominated particles at T1, T2, and T3 located in (A). Reproduced under Create Commons License 4.0 (CC-BY) from reference [110], © The Author(s), 2020.

imaging technique described in Chapter 2 for water extracted particles (Figure 3.16a) and those collected on the impactor (Figure 3.16b). The fluorescence lifetime of the dye sensitive to changes in viscosity is calibrated to generate the data in Figure 3.16d, which shows a significant increase in the viscosity of myrcene SOA particles, with decreasing RH. These images also allowed for monitoring the dynamics (i.e., diffusion time) of individual particle hydration in real time to obtain a rough estimate of the effective diffusion coefficient of the myrcene droplets to be 1.5×10^{-10} cm^2 s^{-1} [118].

The dependency of aerosol morphologies on size were demonstrated in a recent study by Lee et al. [119] using SSA. Figure 3.17 shows the SEM images of morphological categories in A and their size dependency in B, based on the particle cutoff size ranges from micro-orifice uniform deposit impactor (MOUDI). The trend in the morphologies with decreasing size shown in Figure 3.17B can be described as follows: prism-like, core–shell, and rounded morphologies accounted for more than 99% of the entire SSA population, and the rod-inclusion and aggregate morphologies had their greatest abundance in the largest size range (1.0–1.8 µm) at 0.5 and 4.1%, respectively, which decreased with decreasing particle size. Analysis of the relative abundance of the three main morphologies, prism-like, core–shell, and rounded morphologies, with decreasing particle size was correlated with the chemical composition. It was found that decreasing particle size resulted in a significant increase in the rounded and core–shell morphologies, at the expense of the relative abundance of prism-like morphology (which is larger, composed of the inorganic fraction). Concurrently, decreasing particle size resulted in a significant increase in the organic mass fraction of SSA.

3.4 Aerosol optical properties

The term aerosol optical properties refers to parameters that quantify the ability of aerosols to scatter and absorb shortwave and longwave radiation through the imaginary (k) and real (n) parts of the particle refractive index. The refractive index of scattering particles is used in Mie theory calculations along with size distribution to compare calculated phase function and linear polarization for light scattering with experimental ones [120]. Mie theory was derived for spherical particles and, hence, it is limited in predicting particle scattering from nonspherical and irregularly-shaped particles, as illustrated in the case of mineral dust and its components [121]. Details on the measurements of optical properties are provided in Chapter 2. These parameters are used in climate models for the calculations of the radiative forcing due to aerosol-radiation interactions (RFari). Figure 1.2 shows the annual mean top-of-the-atmosphere radiative forcing due to aerosol–radiation interactions (RFari, in W m^{-2}) due to different anthropogenic aerosol types, for the 1750–2010 period [122]. Values of RFari clearly show variation, depending on the chemical composition. These properties are not static and change during the lifetime of the aerosols.

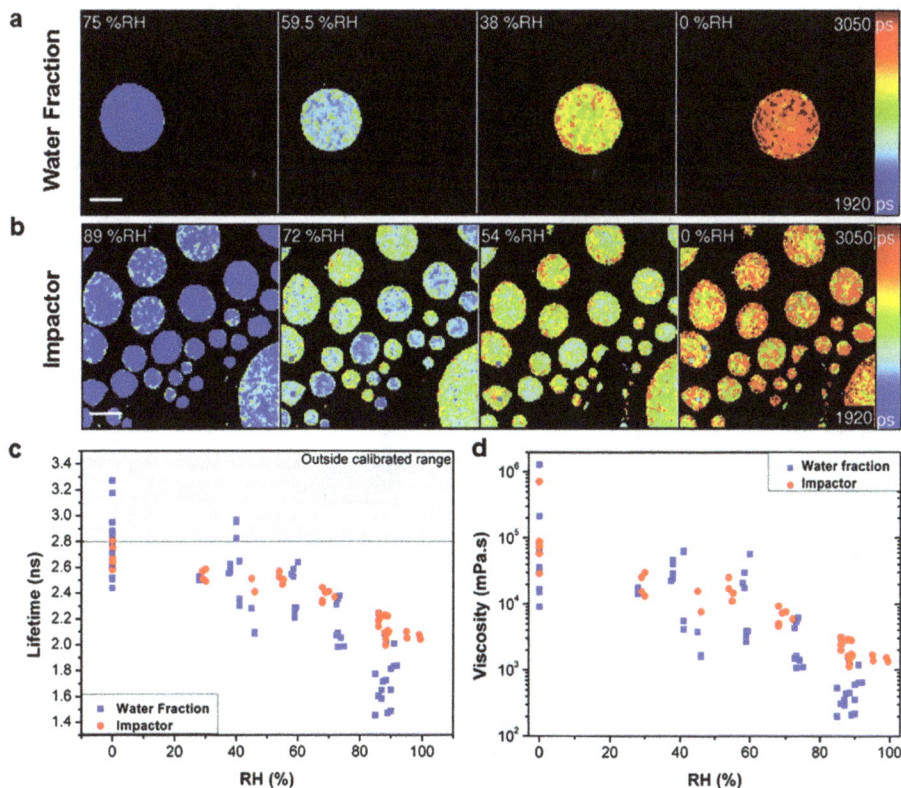

Figure 3.16: Measuring viscosity of oxidized myrcene aerosol during RH change using fluorescence lifetime imaging (FLIM) of viscosity technique. Images of (a) water extracted fraction and (b) impactor collected samples at decreasing RH (scale bar 40 μm). Images show the increase of fluorescence lifetime, with decreasing RH. (c) Averaged fluorescence lifetimes of individual aerosol particles that were either water extracted (blue) or impactor collected (red), measured at various RHs. (d) Lifetimes converted into viscosity according to the calibration in Fig. S2, in reference [118]. The lifetime values above the threshold in (c) (gray shaded area) are outside the calibrated range for the molecular rotor and were not converted in (d). Reproduced under Create Commons License 3.0 (CC-BY) from reference [118], ©The Authors, 2016.

This section aims to highlight recent studies on the optical properties of different aerosol types, with a focus on those with dominant light absorption properties, since it results in positive RFari and, hence, have direct climate relevance [123], namely mineral dust, black carbon (BC), and brown carbon (BrC). Figure 3.18 shows a graphical representation for the absorption and scattering optical parameters, namely k, mass absorption coefficient (MAC), and single scattering albedo (SSA) for these three aerosol types. Conceptually, the net shortwave aerosol absorption, usually quantified through the absorbing aerosol optical depth (AAOD) can, therefore, be thought of as the sum of the contributions of the above three separate

(A)

(B)

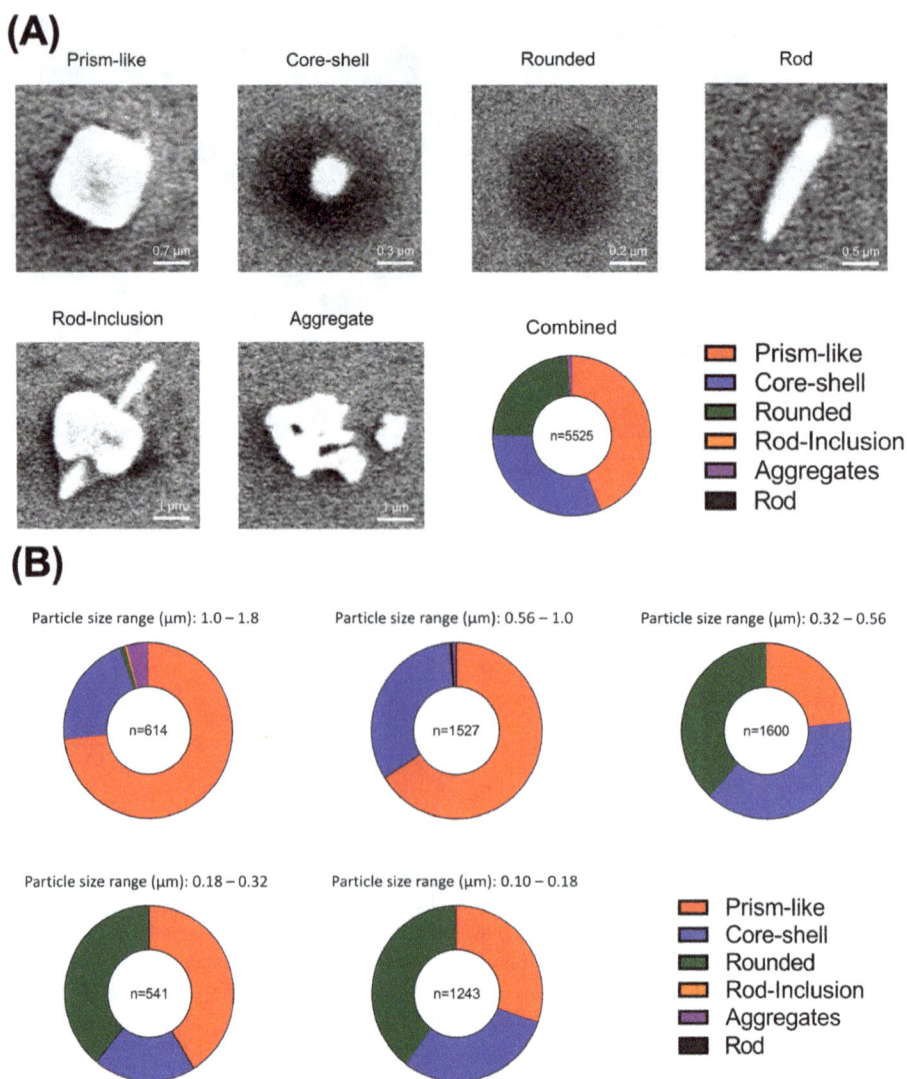

Figure 3.17: (A) Selected illustrative scanning electron micrographs of six morphological categories identified for nascent sea spray aerosols. A combined total of 5,525 individual particles were characterized, and the percent breakdown for each morphology is shown in the colored pie chart. (B) Distribution of the morphological categories based on the particle cutoff size ranges from MOUDI stages 5 (largest, top left) to 9 (smallest, bottom middle) at 80% RH. Reprinted (adapted) with permission from reference [119], © American Chemical Society, 2020.

species, integrated over the atmospheric column [123]. Modeling AAOD is challenging and, hence, model validation, because the observed aerosol mixing state makes clear distinctions between separate aerosol types difficult [111, 123]. The SSA values

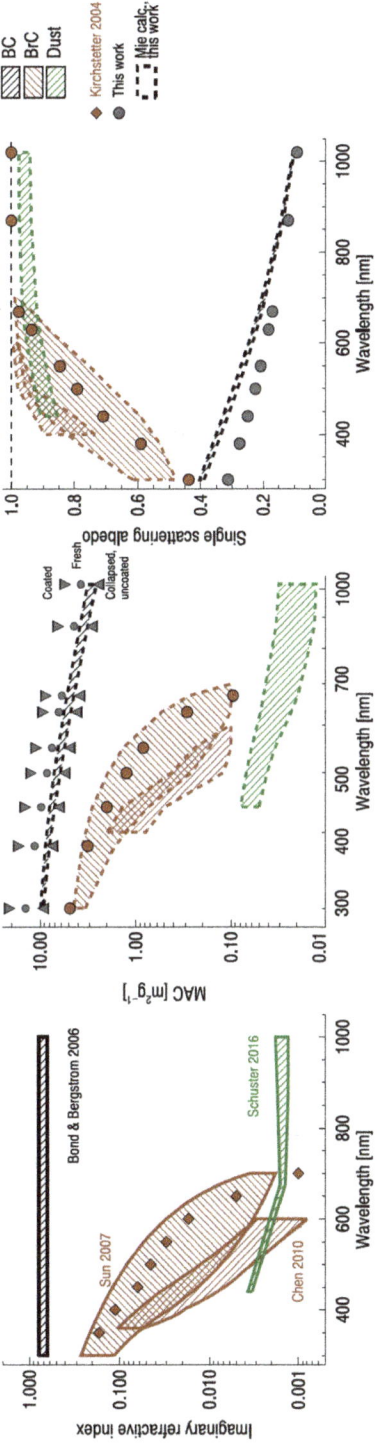

Figure 3.18: Physical parameters used in modeling aerosol absorption: wavelength dependence of the imaginary refractive index (left), MAC (middle), and SSA (right). The term "this work" refers to reference [123]. Reproduced under Create Commons License 4.0 (CC-BY) from reference [123].
© The Authors, 2018.

range 0–1 for purely scattering. Aerosols that have SSA values below ~0.85–0.9 indicate absorption of solar radiation and hence, have a net positive RF [124].

3.4.1 Mineral dust

Mineral dust in the atmosphere is the most abundant atmospheric aerosol by mass
in most global aerosol models [125, 126], and, as a result, its absorption of solar radiation can dominate that of black and brown carbon in some regions and seasons
[127, 128]. As climate forcing is highly sensitive to the optical properties of dust particles [129], dust radiative forcing can be positive (heating) or negative (cooling),
depending on the values of key variables that include height of dust layer, particle
size, and aerosol optical depth (AOD) [130]. Although dust particles weakly absorb
visible radiation, SSA of dust particles is typically in excess of 0.95 at the peak of
the solar spectrum [131, 132]. The SSA value depends on the particle size and mineralogy (particularly the hematite and goethite content). Scanza et al. [133] reported
the simulation of dust radiative forcing as a function of both mineral composition
and size at the global scale, using mineral soil maps for estimating emissions. Previous field measurements of Asian outflows showed that volatiles efficiently transfer
onto dust particles, [134], thus creating an environment for possible heterogenous
reactions. Light-absorbing inclusions on dust have strong effect on its optical properties. For example, internal mixing of dust particles with strongly light-absorbing
black carbon has been shown to decrease SSA for smaller particle size [135].

Al-Abadleh and coworkers reported a chemical process that made dust more
light absorbing [101]. Using ATD and hematite nanoparticles as laboratory standards
and proxies for hematite-rich natural dust, respectively, they showed that reactions
with catechol over 14 days of simulated acidic atmospheric aging (pH 1 and 3) led to
the formation of black polycatechol, increasing the particles' ability to absorb light
over a wide range of wavelengths. This simulated aging process was is in contrast with
the reactivity of oxalate and sulfate, which form complexes with iron sites, promoting
the dissolution of iron minerals in dust. Hence, this aging chemistry with catechol may
change the radiative forcing of dust aerosol from negative to positive, similar to that of
black carbon. Mixing with pollution emissions during mineral dust transport is also
recognized to alter the optical properties of dust [110, 136, 137]. Bi et al. [137] quantified
optical parameters of pure and transported anthropogenic dust over East and Central
Asia, for improving accuracy of remote sensing applications and global climate
models.

3.4.2 Black carbon (BC or soot)

Black carbon (BC) or soot is the most abundant light-absorbing carbonaceous aerosol type, which results from combustion of fossil fuel and biofuel. Within the current uncertainty, BC has a net positive radiative forcing, second only to CO_2 [138, 139]. Bond and Bergstrom [140] reviewed the literature on the relationship between the structure of BC and its optical parameters to be used in models. Flame-generated BC contains sp^3 bonds (as in diamond) and sp^2 bonds with loosely held valence electrons in π-orbitals (as in graphite), in addition to hydrogen and oxygen. Bond and Bergstrom [140] suggest a narrow range of refractive indices, as highlighted in Figure 3.18 (left), and a mass-normalized absorption cross section of 7.5 ± 1.2 m^2 g^{-1} at 550 nm, for uncoated particles (Figure 3.18 (middle)).

Freshly-emitted BC has "fractal" structure, where nanometer size particles form conglomerates. This material absorbs electromagnetic radiation across a broad spectrum, because the energy levels of the loosely-held electrons are closely spaced. As highlighted above, BC can be externally or internally mixed within an aerosol population, hence affecting the overall morphology of the particles. Terms that highlight morphological changes of BC particles during atmospheric aging include "encapsuled", "inclusions", "collapsed", "uncoated", and "coated". Figure 3.19 show changes in morphology, with atmospheric aging time leading to enhancement in MAC and direct radiative forcing (DRF) values, based on the experimental evidence by Peng et al. [141]. MAC values for aged BC are higher than those of fresh/uncoated particles (Figure 3.18 middle). In their models [142], Boucher et al. [142] used average values of 11 m^2 g^{-1} for aged BC particles and stated that converting enhancements to increased DRF is still work in progress. These enhancements in MAC have been explained by an optical lensing effect due to the coatings. In their review, Samset et al. [123] provided a summary table for enhancement in absorption (E_{abs}) as a function of wavelength for three different aging cases studied in the literature: "Pure → Fresh", "Fresh → Aged", and "Pure → Aged", where "Pure" refers to uncoated and collapsed BC, "Fresh" refers to freshly emitted BC, while "Aged" refers to aged BC that has become coated. Table 3.6 lists these E_{abs} values and the source studies. Zhang et al. [143] correlated E_{abs} values with chemical composition from field data collected over four seasons and found that the significant wavelength-dependent E_{abs} increase (from 1 to 2) is correlated with the aerosol photochemical aging associated with the production of highly oxidized SOA, especially at summertime. Other factors that contribute to the uncertainty in BC radiative forcing calculations are residence time, vertical profiles, and emission inventories [123].

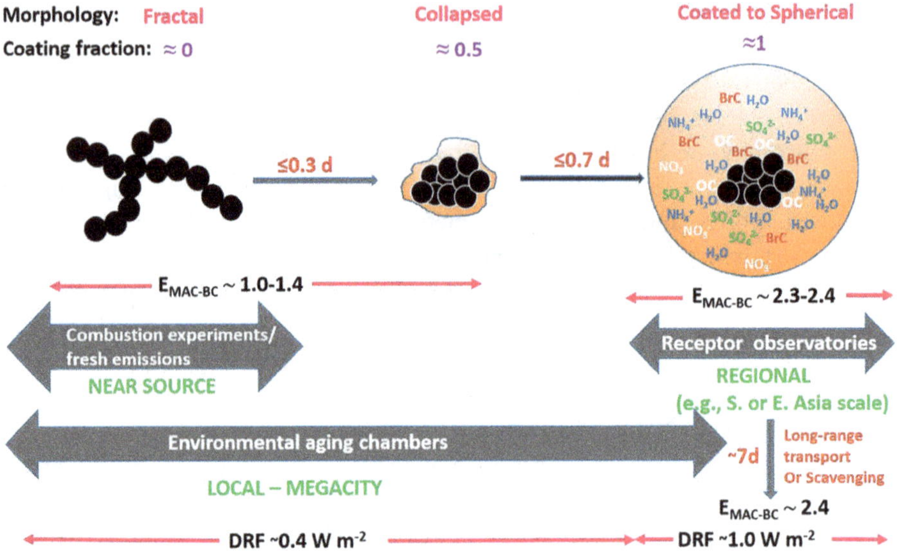

Figure 3.19: Time–course evolution of BC aerosol composition, light absorption (where E_{MAC-BC} is the enhancement because of coatings relative to fresh/uncoated particles), and associated climate effects (as DRF). Reproduced Create Commons License (CC BY-NC-ND 4.0) from reference [144]. © The Authors, 2016.

Table 3.6: Overview of enhancement factors from a number of studies.

Reference	$E_{abs,fresh}$	$E_{abs,aged}$	$E_{abs,total}$	λ (nm)
	Pure → fresh	Fresh → aged	Pure → aged	
Bond and Bergstrøm (2006) [145]	1.5	1.5	(2.3)	
Cappa et al. (2012) [146]	1.06	1.2		532
Cui et al. (2016) [147]	1.4	1.7	3	678
Peng et al. (2016) [148]			2.4	532
Liu et al. (2015) [149]	1.1	1.4		781
Healy et al. (2015) [150]	1.0			781
Nakayama et al. (2014) [151]	1.1			781
Sinha et al. (2017) [152]			1.44	565
Lan et al. (2013) [153]	1.07			532
Liu et al. (2017) [154]		1.1–1.6		405, 532, 781

Reproduced under Create Commons License 4.0 (CC-BY) from reference [123], © The Authors, 2018.

3.4.3 Brown carbon (BrC)

The solar light absorbing fraction of organic aerosol from primary (mainly, combustion) and secondary sources is called brown carbon (BrC), which is structurally differs from BC, resulting in strong absorption at short wavelengths (below 500 nm), with some classes of BrC showing absorption up to 600 nm [123, 155–157]. Figure 3.18 shows the optical parameters and the level of uncertainty associated with each parameter used in predicting the RF of BrC. Factors contributing to this relatively large variability include structure and effects of atmospheric aging, such as lensing and photobleaching, and emission inventory. The molecular structure of BrC includes tar balls, humic-like substances (HULIS), and water-soluble organic carbon (WSOC) with different classes of chromophores such as aromatic carboxylic acids, nitro-phenols; substituted, heterocyclic, and pure polycyclic aromatic hydrocarbons [123, 155, 156, 158, 159]. These structural differences result in variable polarities, volatility [160], viscosity [161], and hygroscopicity [162], which affect reactivity. Chemical processes that lead to the formation of BrC are discussed in Chapter 4.

The lensing effect that increases absorption of BrC was found to be larger in internally mixed aerosol with BC than with non-light-absorbing organic carbon [156]. On the other hand, in laboratory studies, photobleaching due to irradiation with UV light at 365 nm of low molecular weight BrC was found to occur faster than with high molecular weight [156]. Photobleaching is a photolysis and oxidation process that breaks C=C and C-N bonds contributing to the light absorption properties of BrC. Simulating photobleaching of BrC aerosol under realistic conditions using sunlight as the source remains challenging. Li et al. [163] conducted lab experiments on the optical and chemical transformations of biomass burning (wood) tar proxy aerosols by nitrate radicals ($NO_3^•$) and ozone oxidation in the dark, followed by photolysis and photochemical $^•OH$ reactions in simulated daytime. The summary of their results is shown in Figure 3.20. They found that $NO_3^•$ reactions form secondary chromophores, such as nitro aromatic compounds and organonitrates, causing an increase in MAC of the aerosols by a factor of 2–3. Subsequent $^•OH$ oxidation and direct photolysis, both, decompose chromophores in the $NO_3^•$-aged wood tar aerosols, thus decreasing absorption. Studies by Fleming et al. [164] showed that photodegradation lifetime of BrC depends on the fuel type, which ranges from 3.4 ± 1 for Subalpine fir to 14 ± 1 for Juniper.

In an attempt at reducing uncertainties associated with predicting RF [165] of BrC, Lu et al. [166] developed a method to constrain the BrC absorptivity at the emission inventory level, using laboratory and field observations of primary organic aerosol (POA). The main output was wavelength-dependent imaginary refractive indices (k_{OA}) for different fuel types that include biomass/biofuel, lignite, propane, and oil. The authors acknowledged that effects of mixing, aging, and SOA formation need to be added to models that estimate the radiative effect of BrC.

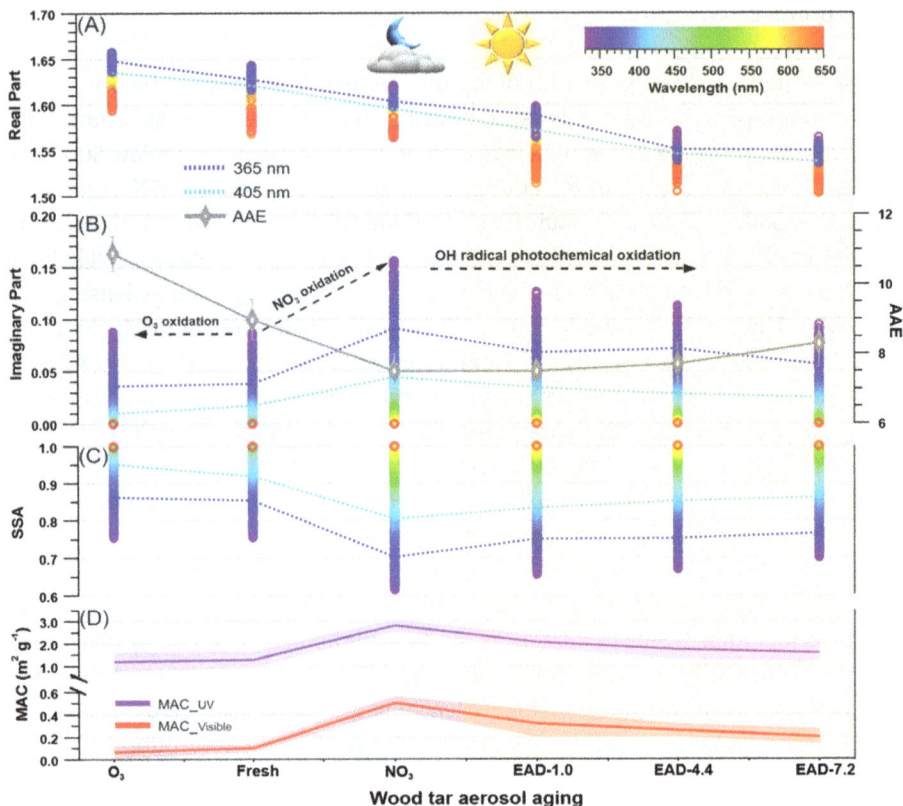

Figure 3.20: Optical evolution of wood tar aerosols through various atmospheric aging processes, including O_3 and $NO_3^•$ reactions in the dark (gray area: nighttime aging simulation) and photochemical transformation of $NO_3^•$-aged wood tar aerosols (yellow area: daytime photochemical simulation). (A) Continuous wavelength-dependent complex real refractive index (RI) as a function of atmospheric transformation. (B) Continuous wavelength-dependent complex imaginary RI as a function of atmospheric evolution (left y-axis). Absorption Ångström exponent (AAE) was derived to indicate the wavelength dependence of light absorption (right y-axis). (C) Wavelength-dependent SSA was estimated, based on retrieved refractive indices (RIs) for wood tar particles at 200 nm. (D) Changes of the wavelength-weighted mean MAC ($m^2 g^{-1}$) for wood tar aerosols via various atmospheric aging processes. Two categories are shown: near-UV absorption (330–400 nm) and visible absorption (400–550 nm). Uncertainties for each RI and SSA distribution (±0.005 for the real part, ±0.006 for the imaginary part, and ±0.013 for the SSA on average from 330 to 550 nm) are not presented for clarity. The term "EAD" refers to the experimental hydroxyl radical concentration, which is equivalent to 1.0–7.2 days of ambient daytime oxidation. EAD-1.0 indicating equivalent to 1.0 day of atmospheric aging and so forth. Reproduced under Create Commons License (CC BY) from reference [163], © American Chemical Society, 2020.

References

[1]   Wu Z, Chen J, Wang Y, Zhu Y, Liu Y, Yao B, et al. Interactions between water vapor and atmospheric aerosols have key roles in air quality and climate change. Natl Sci Rev. 2018; 5(4):452–454.

[2]   Vu TV, Delagad-Saborit JM, Harrison RM. A review of hygroscopic growth factors of submicron aerosols from different sources and its implication for calculation of lung deposition efficiency of ambient aerosols. Air Qual Atmos Health. 2015;8:429–440.

[3]   Farmer DK, Cappa CD, Kreidenweis SM. Atmospheric processes and their controlling influence on cloud condensation nuclei activity. Chem Rev. 2015;115(10):4199–4217.

[4]   Usher CR, Michel AE, Grassian VH. Reactions on Mineral Dust. Chem Rev. 2003;103:4883–4940.

[5]   Tang M, Cziczo DJ, Grassian VH. Interactions of water with mineral dust aerosol: water adsorption, hygroscopicity, cloud condensation, and ice nucleation. Chem Rev. 2016;116:4205–4259.

[6]   Stumm W. Chemistry of the Solid-Water Interface. New York: John Wiely & Sons, Inc.; 1992.

[7]   Tang M, Zhang H, Gu W, Gao J, Jian X, Shi G, et al. Hygroscopic properties of saline mineral dust from different regions in china: geographical variations, compositional dependence, and atmospheric implications. J Geophys Res-Atmos. 2019;124:10844–10857.

[8]   Gaston CJ, Pratt KA, Suski KJ, May NW, Gill TE, Prather KA. Laboratory studies of the cloud droplet activation properties and corresponding chemistry of saline playa dust. Environ Sci Technol. 2017;51:1348–1356.

[9]   Goodman AL, Bernard ET, Grassian VH. Spectroscopic study of nitric acid and water adsorption on oxide particles: Enhanced nitric acid uptake kinetics in the presence of adsorbed water. J Phys Chem A. 2001;105(26):6443–6457.

[10]   Atkins PW, de Paula J. Physical Chemistry. 7th edn, New York: W. H. Freeman and Company; 2008.

[11]   Chen L, Peng C, Gu W, Fu H, Jian X, Zhang H, et al. On mineral dust aerosol hygroscopicity. Atmos Chem Phys. 2020;20:13611–13626.

[12]   Hatch CD, Wiese JS, Crane C, Harris KJ, Kloss HG, Baltrusaitis J. Water adsorption on clay minerals as a function of relative humidity: application of BET and Freundlich adsorption models. Langmuir. 2012;28(3):1790–1803.

[13]   Mirrielees JA, Brooks SD. Instrument artifacts lead to uncertainties in parameterizations of cloud condensation nucleation. Atmos Meas Tech. 2018;11:6389–6407.

[14]   Baltrusaitis J, Grassian VH. Calcite (10-14) surface in humid environments. Surf Sci. 2009;603:L99–L104.

[15]   Schuttlefield J, Al-Hosney H, Zachariah A, Grassian VH. Attenuated total reflection Fourier transform infrared spectroscopy to investigate water uptake and phase transitions in atmospherically relevant particles. App Spectrosc. 2007;61:283–292.

[16]   Ma Q, He H, Liu Y. In situ DRIFTS study of hygroscopic behavior of mineral aerosol. J Environ Sci. 2010;22:555–560.

[17]   Keskinen H, Romakkaniemi S, Jaatinen A, Miettinen P, Saukko E, Jorma J, et al. On-line characterization of morphology and water adsorption on fumed silica nanoparticles. Aerosol Sci Technol. 2011;45(12):1441–1447.

[18]   Kumar P, Sokolik IN, Nenes A. Measurements of cloud condensation nuclei activity and droplet activation kinetics of fresh unprocessed regional dust samples and minerals. Atmos Chem Phys. 2011;11(7):3527–3541.

[19]   Dalirian M, Keskinen H, Ahlm L, Ylisirnioe A, Romakkaniemi S, Laaksonen A, et al. CCN activation of fumed silica aerosols mixed with soluble pollutants. Atmos Chem Phys. 2015; 15(7):3815–3829.

[20] Hung HM, Wang KC, Chen JP. Adsorption of nitrogen and water vapor by insoluble particles and the implication on cloud condensation nuclei activity. J Aerosol Sci. 2015;86:24–31.

[21] Gustafsson RJ, Orlov A, Badger CL, Griffiths PT, Cox RA, Lambert RM. A comprehensive evaluation of water uptake on atmospherically relevant mineral surfaces: DRIFT spectroscopy, thermogravimetric analysis and aerosol growth measurement. Atmos Chem Phys. 2005;5(12):3415–3421.

[22] Kumar P, Sokolik IN, Nenes A. Cloud condensation nuclei activity and droplet activation kinetics of wet processed regional dust samples and minerals. Atmos Chem Phys. 2011;11 (16):8661–8676.

[23] Herich H, Tritscher T, Wiacek A, Gysel M, Weingartner E, Lohmann U, et al. Water uptake of clay and desert dust aerosol particles at sub- and supersaturated water vapor conditions. Phys Chem Chem Phys. 2009;11(36):7804–7809.

[24] Garimella S, Huang Y-W, Seewald JS, Cziczo DJ. Cloud condensation nucleus activity comparison of dry- and wet-generated mineral dust aerosol: The significance of soluble material. Atmos Chem Phys. 2014;14(12):6003–6019.

[25] Joshi N, Romanias MN, Riffault V, Thevenet F. Investigating water adsorption onto natural mineral dust particles: Linking DRIFTS experiments and BET theory. Aeolian Res. 2017;27:35–45.

[26] Ibrahim S, Romanias MN, Alleman LY, Zeineddine MN, Angeli GK, Trikalitis PN, et al. Water interaction with mineral dust aerosol: particle size and hygroscopic properties of dust. ACS Earth Space Chem. 2018;2(4):376–386.

[27] White SC, Case ED. Characterization of fly ash from coal-fired power plants. J Mater Sci. 1990;25:5215–5219.

[28] Srinivasachar S, Helble JJ. Mineral behaviour during coal combustion 1. Pyrite transformations. Prog Energy Combust Sci. 1990;16:281–292.

[29] Peng C, Gu W, Li R, Lin Q, Ma Q, Jia S, et al. Large variations in hygroscopic properties of unconventional mineral dust. ACS Earth Space Chem. 2020;4(10):1823–1830.

[30] Chen H, Laskin A, Baltrusaitis J, Gorski CA, Scherer MM, Grassian VH. Coal fly ash as a source of iron in atmospheric dust. Environ Sci Technol. 2012;46:2112–2120.

[31] Navea JG, Richmond GL, Stortini T, Greenspan J. Water adsorption isotherms on fly ash from several sources. Langmuir. 2017;33:10161–10171.

[32] Borgatta J, Paskavitz A, Kim D, Navea JG. Comparative evaluation of iron leach from fly ash from different source regions under atmospherically relevant conditions. Environ Chem. 2016;13:902–912.

[33] Kuang Y, Xu W, Tao J, Ma N, Zhao C, Shao M. A review on laboratory studies and field measurements of atmospheric organic aerosol hygroscopicity and its parameterization based on oxidation levels. Current Poll Rep. 2020;6:410–424.

[34] Pajunoja A, Lambe AT, Hakala J, Rastak N, Cummings MJ, Brogan JF, et al. Adsorptive uptake of water by semisolid secondary organic aerosols. Geophys Res Lett. 2015;42:3063–3068.

[35] Rastak N, Pajunoja A, Acosta Navaaro JC, Ma J, Partridge DG, Kirkevag A, et al. Microphysical explanation of the RH-dependent water affinity of biogenic organic aerosol and its importance for climate. Geophys Res Lett. 2017;44:5167–5177.

[36] Mayer JK, Wang X, Santander MV, Mitts BA, Sauer J, Sultana CM, et al. Secondary marine aerosol plays a dominant role over primary sea spray aerosol in cloud formation. ACS Cent Sci. 2020;6(12):2259–2266.

[37] Slikboer S, Grandy L, Blair SL, Nizkorodov SA, Smith RW, Al-Abadleh HA. Formation of light absorbing soluble secondary organics and insoluble polymeric particles from the dark reaction of catechol and guaiacol with Fe(III). Environ Sci Technol. 2015;49(13):7793–7801.

[38] Tran A, William G, Younus S, Ali NN, Blair SL, Nizkorodov SA, et al. Efficient formation of light-absorbing polymeric nanoparticles from the reaction of soluble Fe(III) with C4 and C6 dicarboxylic acids. Environ Sci Technol. 2017;51(17):9700–9708.

[39] Rahman MA, Al-Abadleh HA. Surface water structure and hygroscopic properties of light absorbing secondary organic polymers of atmospheric relevance. ACS Omega. 2018;3: 15519–15529.

[40] Kollman PA, Allen LC. Theory of the hydrogen bond. Chem Rev. 1972;72(3):283–303.

[41] Physical Chemistry of Environmental Interfaces. In: Fairbrother H, Geiger FM, Grassian VH, Hemminger JC, editors; J Phys Chem C. 2009; 113(6):2035–2646.

[42] Physical Chemistry of Aerosols. In: Signorell R, Bertram AK, editors. Phys Chem Chem Phys. 2009;11:7741–8104.

[43] Al-Abadleh HA, Grassian VH. FT-IR study of water adsorption on aluminum oxide surfaces. Langmuir. 2003;19:341–347.

[44] Ewing GE. Ambient thin film water on insulator surfaces. Chem Rev. 2006;106:1511–1526.

[45] Cantrell W, Ewing GE. Thin film water on muscovite mica. J Phys Chem B. 2001;105:5434–5439.

[46] Bagchi B. Water dynamics in the hydration layer around proteins and micelles. Chem Rev. 2005;105(9):3197–3219.

[47] Winter N, Vieceli J, Benjamin I. Hydrogen-bond structure and dynamics at the interface between water and carboxylic acid-functionalized self-assembled monolayers. J Phys Chem B. 2008;112(2):227–231.

[48] Falk M, Ford TA. Infrared spectrum and structure of liquid water. Can J Chem. 1966;44 (14):1699–1707.

[49] Thiel PA, Madey TE. The interaction of water with solid surfaces: Fundamental aspects. Surf Sci Rep. 1987;7(6-8):211–385.

[50] Jeffrey G. An Introduction to Hydrogen Bonding. New York: Oxford University Press; 1997.

[51] Scatena LF, Brown MG, Richmond GL. Water at hydrophobic surfaces: Weak hydrogen bonding and strong orientation effects. Science. 2001;292(5518):908–912.

[52] Shen YR, Ostroverkhov V. Sum-frequency vibrational spectroscopy on water interfaces: Polar orientation of water molecules at interfaces. Chem Rev. 2006;106:1140–1154.

[53] Yesilbas M, Boily J-F. Particle size controls on water adsorption and condensation regimes at mineral surfaces. Sci Rep. 2016;6(32136):1–10.

[54] Cowen S, Al-Abadleh HA. DRIFTS studies on the photodegradation of tannic acid as a model for HULIS in atmospheric aerosols. Phys Chem Chem Phys. 2009;11(3):7838–7847.

[55] Tang M, Gu W, Ma Q, Li YJ, Zhong C, Li S, et al. Water adsorption and hygroscopic growth of six anemophilous pollen species: The effect of temperature. Atmos Chem Phys. 2019;19:2247–2258.

[56] Asay DB, Barnette AL, Kim SH. Effects of surface chemistry on structure and thermodynamics of water layers at solid-vapor interfaces. J Phys Chem C. 2009;113:2128–2133.

[57] Moussa SG, McIntire TM, Szori M, Roeselová M, Tobias DJ, Grimm RL, et al. Experimental and theoretical characterization of adsorbed water on self-assembled monolayers: Understanding the interaction of water with atmospherically relevant surfaces. J Phys Chem A. 2009;113(10):2060–2069.

[58] Lappi SE, Smith B, Franzen S. Infrared spectra of H2160, H2180 and D2O in the liquid phase by single-pass attenuated total internal reflection spectroscopy. Spectrochim Acta A. 2004;60:2611–2619.

[59] Wyss HR, Falk M. Infrared spectrum of HDO in water and in NaCl solution. Can J Chem. 1970;48:607–614.

[60] Raymond EA, Richmond GL. Probing the molecular structure and bonding of the surface of aqueous salt solutions. J Phys Chem B. 2004;108:5051–5059.

[61] Liu D, Ma G, Levering LM, Allen HC. Vibrational spectroscopy of aqueous sodium halide solutions and air-liquid interfaces: observation of increased interfacial depth. J Phys Chem B. 2004;108:2252–2260.

[62] Boily J-F, Fu L, Tuladhar A, Lu Z, Legg BA, Wang ZM, et al. Hydrogen bonding and molecular orientations across thin water films on sapphire. J Coll Inter Sci. 2019;555:810–817.

[63] Tuladhar A, Dewan S, Kubicki JD, Borguet E. Spectroscopy and ultrafast vibrational dynamics of strongly hydrogen bonded OH Species at the α-Al2O3(1120)/H2O Interface. J Phys Chem C. 2016;120:16153–16161.

[64] Al-Abadleh HA, Al-Hosney HA, Grassian VH. Oxide and carbonate surfaces as environmental interfaces: Importance of water in surface composition and surface reactivity. J Molec Catal A. 2005;228:47–54.

[65] Cheng W, Hanna K, Boily J-F. Water vapor binding on organic matter-coated minerals. Environ Sci Technol. 2019;53:1252–1257.

[66] Hemminger JC. Heterogeneous chemistry in the troposphere: A modern surface chemistry approach to the study of fundamental processes. Int Rev Phys Chem. 1999;18(3):387–417.

[67] Orlando F, Waldner A, Bartels-Rausch T, Birrer M, Kato S, Lee M-T, et al. The environmental photochemistry of oxide surfaces and the nature of frozen salt solutions: A new in Situ XPS approach. Top Catal. 2016;59:591–604.

[68] Al-Abadleh HA, Grassian VH. Oxide surfaces as environmental interfaces. Surf Sci Rep. 2003;52:63–162.

[69] Salmeron M. Ambient pressure photoelectron spectroscopy: A new tool for surface science and nanotechnology. Surf Sci Rep. 2008;63:169–199.

[70] Yamamoto M, Bluhm H, Andersson K, Ketteler G, Ogasawara H, Salmeron M, et al. In situ x-ray photoelectron spectroscopy studies of water on metals and oxides at ambient conditions. J Phys: Condes Matter. 2008;20(184025):1–14.

[71] Cavalleri M, Ogasawara H, Petterson LGM, Nilsson A. The interpretation of X-ray absorption spectra of water and ice. Chem Phys Lett. 2002;364:363–370.

[72] Yamamoto S, Kendelewicz T, Newberg JT, Ketteler G, Starr DE, Mysak ER, et al. Water adsorption on alpha-Fe2O3(0001) at near ambient conditions. J Phys Chem C. 2010;114 (5):2256–2266.

[73] Dupin J-C, Gonbeau D, Vinatier P, Levasseur A. Systematic XPS studies of metal oxides, hydroxides and peroxides. Phys Chem Chem Phys. 2000;2:1319–1324.

[74] Wernet P, Nordlund D, Bergmann U, Cavalleri M, Odelius M, Ogasawara H, et al. The structure of the first coordination shell in liquid water. Science. 2004;304:995–999.

[75] Naslund LA, Luning J, Ufuktepe Y, Ogasawara H, Wernet P, Bergmann U, et al. X-ray absorption spectroscopy measurements of liquid water. J Phys Chem B. 2005;109:13835–13839.

[76] Cappa CD, Smith JD, Wilson KR, Messer BM, Gilles MK, Cohen RC, et al. Effects of alkali metal halide salts on the hydrogen bond network of liquid water. J Phys Chem B. 2005;109:7046–7052.

[77] Krepelova A, Huthwelker T, Bluhm H, Ammann M. Surface chemical properties of eutectic and frozen Nacl solutions probed by XPS and NEXAFS. Chem Phys Chem. 2010;11:3859–3866.

[78] Zhou G, Huang L. A review of recent advances in computational and experimental analysis of first adsorbed water layer on solid substrate. Mol Simul. 2020;46:1–17.

[79] Xiao C, Shi P, Yan W, Chen L, Qian L, Kim SH. Thickness and structure of adsorbed water layer and effects on adhesion and friction at nanoasperity contact. Colloids Interfaces. 2019;3(55):1–31.

[80] DelloStritto MJ, Kubicki JD, Sofo JO. Effect of ions on H-bond structure and dynamics at the quartz(101)–water interface. Langmuir. 2016;32:11353–11365.

[81] Song X, Boily J-F. Water vapor adsorption on goethite. Environ Sci Technol. 2013;47: 7171–7177.

[82] Boily J-F, Yesilbas M, Md. Musleh Uddin M, Baiqing L, Trushkina Y, Salazar-Alvarez G. Thin water films at multifaceted hematite particle surfaces. Langmuir. 2015;31:13127–13137.

[83] Ranathunga DTS, Shamir A, Dai X, Nielsen SO. Molecular dynamics simulations of water condensation on surfaces with tunable wettability. Langmuir. 2020;36:7383–7391.

[84] Martin ST. Phase transitions of aqueous atmospheric particles. Chem Rev. 2000;100 (9):3403–3454.

[85] Mauer LJ, Taylor LS. Water-solids interactions: Deliquescence. Annu Rev Food Sci Technol. 2010;1:41–63.

[86] Krieger UK, Marcolli C, Reid JP. Exploring the complexity of aerosol particle properties and processes using single particle techniques. Chem Soc Rev. 2012;41:6631–6662.

[87] Bilde M, Barsanti K, Booth M, Cappa CD, Donahue NM, Emanuelsson EU, et al. Saturation vapor pressures and transition enthalpies of low-volatility organic molecules of atmospheric relevance: From dicarboxylic acids to complex mixtures. Chem Rev. 2015;115:4115–4156.

[88] Bertram AK, Martin ST, Hanna SJ, Smith ML, Bodsworth A, Chen Q, et al. Predicting the relative humidities of liquid-liquid phase separation, efflorescence, and deliquescence of mixed particles of ammonium sulfate, organic material, and water using the organic-to-sulfate mass ratio of the particle and the oxygen-to-carbon elemental ratio of the organic component. Atmos Chem Phys. 2011;11:10995–11006.

[89] Karthika S, Radhakrishnan TK, Kalaichelvi P. A review of classical and nonclassical nucleation theories. Crys Growth Des. 2016;16(11):6663–6681.

[90] Attwood AR, Greenslade ME. Deliquescence behavior of internally mixed clay and salt aerosols by optical extinction measurements. JPhys Chem A. 2012;116:4518–4527.

[91] Al-Abadleh HA, Grassian VH. Phase transitions in magnesium nitrate thin films: A transmission FT-IR study of the deliquescence and efflorescence of nitric acid reacted magnesium oxide interfaces. J Phys Chem B. 2003;107:10829–10839.

[92] Kanji ZA, Ladino LA, Wex H, Boose Y, Burkert-Kohn M, Cziczo DJ, et al. Overview of ice nucleating particles. Meteoro Monographs. 2017;58:1–33.

[93] Knopf DA, Alpert PA, Wang B. The role of organic aerosol in atmospheric ice nucleation: A review. ACS Earth Space Chem. 2018;2:168–202.

[94] Murray BJ, O'Sullivan D, Atkinson JD, Webb ME. Ice nucleation by particles immersed in supercooled cloud droplets. Chem Soc Rev. 2012;41:6519–6554.

[95] Lamb D. Cloud Microphysics. In: North GR, Pyle J, Zhang F, editors. Encyclopedia of Atmospheric Sciences. 2nd ed, Elsevier; 2015. p. 133–140.

[96] Umo NS, Wagner R, Ullrich R, Kiselev A, Saathoff H, Weidler PG, et al. Enhanced ice nucleation activity of coal fly ash aerosol particles initiated by ice-filled pores. Atmos Chem Phys. 2019;19:8783–8800.

[97] DeMott PJ, Prenni AJ, McMeeking GR, Sullivan RC, Petters MD, Tobo Y, et al. Integrating laboratory and field data to quantify the immersion freezing ice nucleation activity of mineral dust particles. Atmos Chem Phys. 2015;15:393–409.

[98] Yun J, Link N, Kumar A, Schukarev A, Davidson J, Lam A, et al. Surface composition dependence on the ice nucleating ability of potassium-rich feldspar. ACS Earth Space Chem. 2020;4(6):873–881.

[99] Whale TF, Holden MA, Wilson TW, O'Sullivan D, Murray BJ. The enhancement and suppression of immersion mode heterogeneous ice-nucleation by solutes. Chem Sci. 2018;9:4142–4151.

[100] Chatre C, Emmelin C, Urbaniak S, George C, Phase CC. Transformations of liquid drops containing mineral dust and organic compound (citric acid). Crys Growth Des. 2019;19:4619–4624.

[101] Link N, Removski N, Yun J, Fleming L, Nizkorodov SA, Bertram AK, et al. Dust-catalyzed oxidative polymerization of catechol under acidic conditions and its impacts on ice nucleation efficiency and optical properties. ACS Earth Space Chem. 2020;4(7):1127–1139.

[102] Baustian KJ, Cziczo DJ, Wise ME, Pratt KA, Kulkarni G, Hallar AG, et al. Importance of aerosol composition, mixing state, and morphology for heterogeneous ice nucleation: A combined field and laboratory approach. J Geophys Res. 2012;117:D06217.

[103] Yun Y, Penner JE. An evaluation of the potential radiative forcing and climatic impact of marine organic aerosols as heterogeneous ice nuclei. Geophys Res Lett. 2013;40:4121–4126.

[104] Yesilbas M, Boily J-F. Thin ice films at mineral surfaces. J Phys Chem Lett. 2016;7:2849–2855.

[105] Shiraiwa M, Li Y, Tsimpidi AP, Karydis VA, Berkemeier T, Pandis SN, et al. Global distribution of particle phase state in atmospheric secondary organic aerosols. Nature Comm. 2017;8:1–7.

[106] Wong Derieux WS, Li Y, Lin P, Laskin J, Laskin A, Bertram AK, et al. Predicting the glass transition temperature and viscosity of secondary organic material using molecular composition. Atmos Chem Phys. 2018;18:6331–6351.

[107] Ye Q, Robinson ES, Ding X, Ye P, Sullivan RC, Donahue NM. Mixing of secondary organic aerosols versus relative humidity. Proc Natl Acad Sci USA. 2016;113(45): 12649–12654.

[108] Riemer N, West M. Quantifying aerosol mixing state with entropy and diversity measures. Atmos Chem Phys. 2013;18:12595–12612.

[109] Bondy AL, Bonanno D, Moffet RC, Wang B, Laskin A, Ault AP. The diverse chemical mixing state of aerosol particles in the southeastern United States. Atmos Chem Phys. 2018;18:12595–12612.

[110] Xu L, Fukushima S, Sobanska S, Murata K, Naganuma A, Liu L, et al. Tracing the evolution of morphology and mixing state of soot particles along with the movement of an Asian dust storm. Atmos Chem Phys. 2020;20:14321–14332.

[111] Stevens R, Dastoor A. A review of the representation of aerosol mixing state in atmospheric models. Atmosphere. 2019;10(168), doi:10.3390/atmos10040168.

[112] You Y, Smith ML, Song M, Martin ST, Bertram AK. Liquid–liquid phase separation in atmospherically relevant particles consisting of organic species and inorganic salts. Int Rev Phys Chem. 2014;33(1):43–77.

[113] Li W, Sun J, Xu L, Shi Z, Riemer N, Sun Y, et al. A conceptual framework for mixing structures in individual aerosol particles. J Geophys Res-Atmos. 2016;121:13784–13798, doi:10.1002/2016JD025252.

[114] Song M, Marcolli C, Krieger UK, Lienhard DM, Peter T. Morphologies of mixed organic/inorganic/aqueous aerosol droplets. Faraday Discuss. 2013;165(1):289–316.

[115] Song Y-C, Be AG, Martin ST, Geiger FM, Bertram AK, Thomson RJ, et al. Liquid–liquid phase separation and morphologies in organic particles consisting of a-pinene and b-caryophyllene ozonolysis products and mixtures with commercially available organic compounds. Atmos Chem Phys. 2020;20:11263–11273.

[116] Losey DJ, Parker RG, Freedman MA. pH dependence of liquid–liquid phase separation in organic aerosol. J Phys Chem Lett. 2016;7:3861–3865.

[117] Losey DJ, Ott E-JE, Freedman MA. Effects of high acidity on phase transitions of an organic aerosol. J Phys Chem A. 2018;122:3819–3828.

[118] Hosny NA, Fitzgerald C, Vysniauskas A, Athanasiadis A, Berkemeier T, Uygur N, et al. Direct imaging of changes in aerosol particle viscosity upon hydration and chemical aging. Chem Sci. 2016;7:1357–1367.

[119] Lee HD, Wigley S, Lee C, Or VW, Hasenecs ES, Stone EA, et al. Physicochemical mixing state of sea spray aerosols: morphologies exhibit size dependence. ACS Earth Space Chem. 2020;4(9):1604–1611.

[120] Wriedt T. Mie Theory: A Review. In: Hergert W, Wriedt T, editors. The Mie Theory: Basics and Applications. vol. 169, Heidelberg: Springer-Verlag; 2012. p. 53–71.

[121] Curtis DB, Meland B, Aycibiin M, Arnold NP, Grassian VH, Young MA, et al. A laboratory investigation of light scattering from representative components of mineral dust aerosol at a wavelength of 550 nm. J Geophys Res. 2008;113:D08210, doi:10.1029/2007JD009387.

[122] Boucher O, Randall D, Artaxo P, Bretherton C, Feingold G, Forster F, et al. Clouds and Aerosols. In: Stocker TF, Qin D, Plattner G-K, Tignor M, Allen SK, Boschung J, et al editors. Climate Change 2013: The Physical Science Basis Contribution of Working Group I to the Fifth Assessment Report of the Intergovernmental Panel on Climate Change. Cambridge, United Kingdom and New York, NY, USA: Cambridge University Press; 2013. p. 571–657.

[123] Samset BH, Stjern CW, Andrews E, Kahn RA, Myhre G, Schulz M, et al. Aerosol absorption: progress towards global and regional constraints. Curr Clim Change Rep. 2018;4:65–83.

[124] Hansen J, Sato M, Ruedy R. Radiative forcing and climate response. J Geophys Res. 1997; 102(D6):6831–6864.

[125] Textor C, Schulz M, Guibert S, Kinne S, Balkanski Y, Bauer S, et al. Analysis and quantification of the diversities of aerosol life cycles within aerocom. Atmos Chem Phys. 2006;6:1777–1813.

[126] Kok JF, Ridley DA, Zhou Q, Miller RL, Zhao C, Heald CL, et al. Smaller desert dust cooling effect estimated from analysis of dust size and abundance. Nature Geosci. 2017;10(4):274–278.

[127] Caponi L, Formenti P, Massabo D, Di Biagio C, Cazaunau M, Pangui E, et al. Spectral- and size-resolved mass absorption efficiency of mineral dust aerosols in the shortwave spectrum: A simulation chamber study. Atmos Chem Phys. 2017;17(11):7175–7191.

[128] Satheesh SK, Moorthy KK. Radiative effects of natural aerosols: A review. Atmos Environ. 2005;39(11):2089–2110.

[129] Liao H, Seinfeld JH. Radiative forcing by mineral dust aerosols: Sensitivity to key variables. J Geophys Res D. 1998;103(D24):31637–31645.

[130] Choobari OA, Zawar-Reza P, Sturman A. The global distribution of mineral dust and its impacts on the climate system: A review. Atmos Res. 2014;138:152–165.

[131] Kaufman YJ, Tanre D, Dubovik O, Karnieli A, Remer LA. Absorption of sunlight by dust as inferred from satellite and ground-based remote sensing. Geophys Res Lett. 2001;28(8):1479–1482.

[132] Dubovik O, Holben B, Tf E, Smirnov A, Kaufman YJ, King MD, et al. Variability of absorption and optical properties of key aerosol types observed in worldwide locations. J Atmos Sci. 2002;59(3):590–608.

[133] Scanza RA, Mahowald N, Ghan S, Zender CS, Kok JF, Liu X, et al. Modeling dust as component minerals in the Community Atmosphere Model: Development of framework and impact on radiative forcing. Atmos Chem Phys. 2015;15:537–561.

[134] Clarke AD, Shinozuka Y, Kapustin VN, Howell S, Huebert B, Doherty S, et al. Size distributions and mixtures of dust and black carbon aerosol in Asian outflow: Physiochemistry and optical properties. J Geophys Res-Atmos. 2004;109(D15):D15S09, doi:10.1029/2003JD004378.

[135] Scarnato BV, China S, Nielsen K, Mazzoleni C. Perturbations of the optical properties of mineral dust particles by mixing with black carbon: A numerical simulation study. Atmos Chem Phys. 2015;15(12):6913–6928.

[136] Denjean C, Cassola F, Mazzino A, Triquet S, Chevaillier S, Grand N, et al. Size distribution and optical properties of mineral dust aerosols transported in the western Mediterranean. Atmos Chem Phys. 2016;16:1081–1104.

[137] Bi J, Huang J, Holben B, Zhang G. Comparison of key absorption and optical properties between pure and transported anthropogenic dust over East and Central Asia. Atmos Chem Phys. 2016;16:15501–15516.

[138] Ramanathan V, Carmichael G. Global and regional climate changes due to black carbon. Nature Geosci. 2008;1(4):221–227.

[139] Bond TC, Doherty S, Fahey DW, Forster PM, Bernsten T, al. e. Bounding the role of black carbon in the climate system: A scientific assessment. J Geophys Res-Atmos. 2013;118 (11):5380–5552.

[140] Bond T, Bergstrom R. Light absorption by carbonaceous particles: An investigative review. Aerosol Sci Technol. 2006;40(1):27–67.

[141] Peng J, Hu M, Guo S, Du Z, Zheng J, Shang D, et al. Markedly enhanced absorption and direct radiative forcing of black carbon under polluted urban environments. Proc Natl Acad Sci USA. 2016;113(16):4266–4271.

[142] Boucher O, Balkanski Y, Hodnebrog O, Myhre CEL, Myhre G, Quaas J, et al. Jury is still out on the radiative forcing by black carbon. Proc Natl Acad Sci USA. 2016;113(35):E5092–E3.

[143] Zhang Y, Favez O, Canonaco F, Liu D, Mocnik G, Amodeo T, et al. Evidence of major secondary organic aerosol contribution to lensing effect black carbon absorption enhancement. Npj Climate Atmos Sci. 2018;47:1–8.

[144] Gustafsson O, Ramanathan V. Convergence on climate warming by black carbon aerosols. Proc Natl Acad Sci USA. 2016;113(16):4243–4245.

[145] Bond TC, Bergstrom RW. Light absorption by carbonaceous particles: an investigative review. Aerosol Sci Technol. 2006;40(1):27–67.

[146] Cappa CD, Onasch TB, Massoli P, Worsnop DR, Bates TS, Cross ES, et al. Radiative absorption enhancements due to the mixing state of atmospheric black carbon. Science. 2012;337(6098):1078–1081.

[147] Cui X, Wang X, Yang L, Chen B, Chen J, Andersson A, et al. Radiative absorption enhancement from coatings on black carbon aerosols. Sci Total Environ. 2016;551:51–56.

[148] Peng J, Hu M, Guo S, Du Z, Zheng J, Shang D, et al. Markedly enhanced absorption and direct radiative forcing of black carbon under polluted urban environments. Proc National Acad Sci. 2016;113(16):4266–4271.

[149] Liu S, Aiken AC, Gorkowski K, Dubey MK, Cappa CD, Williams LR, et al. Enhanced light absorption by mixed source black and brown carbon particles in UK winter. Nat Commun. 2015;6:8435.

[150] Healy RM, Wang JM, Jeong CH, Lee AKY, Willis MD, Jaroudi E, et al. Light-absorbing properties of ambient black carbon and brown carbon from fossil fuel and biomass burning sources. J Geophys Res: Atmos. 2015;120(13):6619–6633.

[151] Nakayama T, Ikeda Y, Sawada Y, Setoguchi Y, Ogawa S, Kawana K, et al. Properties of light-absorbing aerosols in the Nagoya urban area, Japan, in August 2011 and January 2012: Contributions of brown carbon and lensing effect. J Geophys Res: Atmos. 2014;119 (22):12,721–12,39.

[152] Sinha PR, Kondo Y, Koike M, Ogren JA, Jefferson A, Barrett TE, et al. Evaluation of ground-based black carbon measurements by filter-based photometers at two Arctic sites. J Geophys Res: Atmos. 2017;122(6):3544–3572.

[153] Lan Z-J, Huang X-F, Yu K-Y, Sun T-L, Zeng L-W, Hu M. Light absorption of black carbon aerosol and its enhancement by mixing state in an urban atmosphere in South China. Atmos Environ. 2013;69:118–123.

[154] Liu D, Whitehead J, Alfarra MR, Reyes-Villegas E, Spracklen DV, Reddington CL, et al. Black-carbon absorption enhancement in the atmosphere determined by particle mixing state. Nature Geosci. 2017;10(3):184–188.

[155] Laskin A, Laskin J, Nizkorodov SA. Chemistry of atmospheric brown carbon. Chem Rev. 2015;115(10):4335–4382.

[156] Yan J, Wang X, Gong P, Wang C, Cong Z. Review of brown carbon aerosols: Recent progress and perspectives. Sci Total Environ. 2018;634:1475–1485.

[157] Feng Y, Ramanathan V, Kotamarthi VR. Brown carbon: A significant atmospheric absorber of solar radiation?. Atmos Chem Phys. 2013;13:8607–8621.

[158] Laskin A, Lin P, Laskin J, Fleming L, Nizkorodov SA. Molecular Characterization of Atmospheric Brown Carbon. In: Hunt SW, Laskin A, Nizkorodov SA, editors. Multiphase Environmental Chemistry in the Atmosphere. ACS Symposium Series, 1299. Washington DC: American Chemical Society; 2018. p. 261–274.

[159] Andreae MO, Gelencser A. Black carbon or brown carbon? The nature of light-absorbing carbonaceous aerosols. Atmos Chem Phys. 2006;6(10):3131–3148.

[160] Tasoglou A, Louvaris E, Florou K, Liangou A, Karnezi E, Kaltsonoudis C, et al. Aerosol light absorption and the role of extremely low volatility organic compounds. Atmos Chem Phys. 2020;20:11625–11637.

[161] Fraund M, Bonanno DJ, China S, Pham DQ, Veghte D, Weis J, et al. Optical properties and composition of viscous organic particles found in the Southern great plains. Atmos Chem Phys. 2020;20:11593–11606.

[162] Liu D, He C, Schwarz JP, Wang X. Lifecycle of light-absorbing carbonaceous aerosols in the atmosphere. Npj Climate Atmos Sci. 2020;3(40):1–18.

[163] Li C, He Q, Fang Z, Brown SS, Laskin A, Cohen SR, et al. Laboratory insights into the diel cycle of optical and chemical transformations of biomass burning brown carbon aerosols. Environ Sci Technol. 2020;54:11827–11837.

[164] Fleming LT, Lin P, Roberts JM, Selimovic V, Yokelson R, Laskin J, et al. Molecular composition and photochemical lifetimes of brown carbon chromophores in biomass burning organic aerosol. Atmos Chem Phys. 2020;20:1105–1129.

[165] Lin G, Penner JE, Flanner MG, Sillman S, Xu L, Zhou C. Radiative forcing of organic aerosol in the atmosphere and on snow: Effects of SOA and brown carbon. J Geophys Res-Atmos. 2014;119:7453–7476.

[166] Lu Z, Streets DG, Winijkul E, Yan F, Chen Y, Bond T, et al. Light absorption properties and radiative effects of primary organic aerosol emissions. Environ Sci Technol. 2015;49: 4868–4877.

Chapter 4
Interfacial aerosol chemistry

The initial stages in the aging of the atmospheric aerosol particles take place at the surface of these particles. The relative high surface-to-volume ratio of the atmospheric aerosol particles makes surface reactions more important than bulk reactions. These reactions can lead to changes in the surface and in the bulk chemical composition, and can take place under night- and day-time conditions. The extent and rates of reactions in these systems are influenced by the amount of surface water, which varies with relative humidity (RH) and hygroscopicity of the surfaces. Also, the aerosol surface can act as a catalyst for speeding up reactions that are slow in the gas phase. Given the complexity of the aerosol mixing states and their morphologies, as highlighted in Chapter 3, interfacial chemistry in the atmospheric aerosol particles can take place at the gas/solid, gas/semi-solid, liquid/solid, and gas/liquid interfaces (Figure 1.4). In this chapter, specific examples are presented on each of these cases from the synthesis of relevant literature.

4.1 Importance of interfacial reactions

As stated in Chapter 1, the atmospheric aerosol particles cover a range of sizes, from a few nanometers in the nucleation and Aitken modes to greater than a micron in the course mode [1]. The importance of interfacial reactions increases with decrease in particle size. Finlayson-Pitts calculated the percentage of molecules on the surface of a 1 μm diameter particle with a typical density of 1.2 g cm^{-3}, an average molecular mass of 300 g mol^{-1}, and assuming an area per molecule to be 2×10^{-15} cm^2, to be 1.2%, compared with 12% for a 100 nm diameter particle, and 100% for a 3 nm molecular cluster [2]. Figure 4.1 shows an illustration of the 10x increase in the surface-to-volume (S/V) ratio for a 100 nm diameter particle, compared to a 1 μm diameter particle. However, the 'spherical' shape does not represent atmospheric particles that have complex morphologies and mixing states (see Chapter 2 and 3 for details). The actual S/V ratio might in fact be higher than that estimated for a spherical shape. In addition, surface science studies conducted under vacuum revealed that the degree of bonding unsaturation at the surface due to the presence of defect sites and missing 'neighbors' would increase the surface energy, and hence the surface instability. As a result, the overall reactivity of the particle will be controlled by its surface reactivity with ambient species to lower the surface energy of the particle. For example, metal oxides in mineral dust particles are terminated with hydroxyl groups from the chemisorption of water molecules [3]. Hydroxylation reactions lower the surface energy of metal oxides and turn them amphoteric in character [4–6]. Under humid conditions, characteristic of atmospheric conditions, oxide surfaces have residual amounts of the adsorbed

https://doi.org/10.1515/9781501519376-004

water, in addition to hydroxyl groups. Figure 4.2 shows that these humid conditions are between dry and wet conditions, characteristic of vacuum and aqueous phase cases, respectively [7]. As discussed by Grassian [7], while there is a working conceptual theoretical framework that describes the solid interfaces under dry and wet conditions, there is a lack of molecular level understanding of the nature of surfaces in the middle case, i.e., under humid conditions. As outlined in the following sections, the progress made to date highlights keywords that describe the surfaces under dry, wet, and humid conditions. In the case of liquid droplets, surface tension is one of the

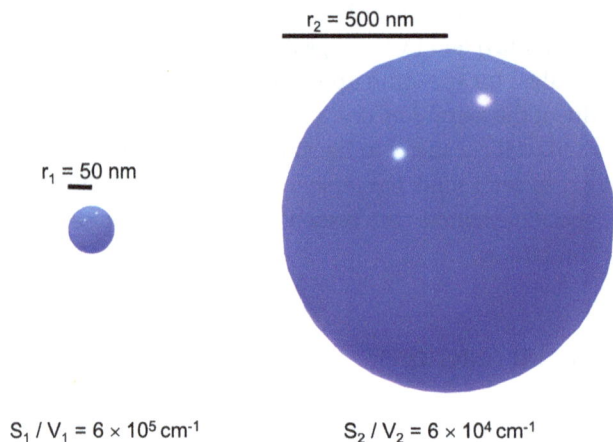

r_2 = 500 nm

r_1 = 50 nm

$S_1 / V_1 = 6 \times 10^5$ cm^{-1}

$S_2 / V_2 = 6 \times 10^4$ cm^{-1}

Figure 4.1: Schematic to illustrate enhancement in surface area-to-volume (S/V) ratio for two spherical particles.

Increasing Water Acidity

Dry Conditions	Humid Conditions	Wet Conditions
Vacuum-Oxide Surface Interface	**Adsorbed Water-Oxide Surface Interface**	**Aqueous-Oxide Surface Interface**
Conceptual Framework: well-defined surfaces, controlled defect site density , oxygen vacancies	Conceptual Framework: defect sites?? Surface coverage, uniform water layers?? pH, pzc??	Conceptual Framework: double layer, solubility, pH, pzc

Figure 4.2: The diagram shows that as the environment goes from dry to humid to wet, i.e., as the water activity increases over several orders of magnitude, the nature of the oxide surfaces under these different environmental conditions changes. There is a clear lack of understanding in the conceptual theoretical framework and other environmental interfaces that helps in understanding the atomic and molecular nature of the oxide surfaces in the presence of increasing water activity changes and of the intermediate regime of humid environments. Figure and caption were reproduced with permission from reference [7], © Elsevier, 2008.

properties that relates the viscosity of droplets and their hygroscopic properties (see Chapters 2 and 3 for details). The following sections provide selected examples on interfacial reactions relevant to atmospheric particles.

4.2 Reactions at the gas/(solid/semi-solid/water) interface

4.2.1 Reactive oxygen species (ROS) and reactive nitrogen species (RNS)

ROS and RNS are classified into radicals and nonradicals. Table 4.1 lists the chemical species in each group. Most of these species can exist in the gas and aqueous phases, and hence contribute to the oxidizing power of the atmosphere and aerosols. The following sections highlight the heterogeneous reactivity of selected ROS and RNS species with components of atmospheric aerosol particles.

Table 4.1: List of ROS and RNS.

Radicals	Nonradicals
Reactive oxygen species (ROS)	
Hydroxyl ($^{\cdot}$OH)	Ozone (O_3)
Hydroperoxyl ($HO_2^{\cdot}$)	Singlet oxygen ($^1O_2$)
Superoxide (O_2^{-})	Hydrogen peroxide (H_2O_2)
Peroxyl ($RO_2^{\cdot}$)	Hypochlorous acid (HOCl)
Alloxyl ($RO^{\cdot}$)	Hypobromous acid (HOBr)
Reactive nitrogen species (RNS)	
Nitric oxide ($NO^{\cdot}$)	Nitrous acid (HONO)
Nitrogen dioxide ($NO_2^{\cdot}$)	Nitric acid (HNO_3)
Nitrate radical ($NO_3^{\cdot}$)	Nitrosyl cation (NO +)
	Nitrosyl anion (NO^{-})
	Peroxynitrous acid (ONOOH)
	Alkyl peroxynitrite (ROONO)
	Dinitrogen pentoxide (N_2O_5)

4.2.2 Ozone (O_3)

4.2.2.1 Sources and sinks of tropospheric ozone

Ozone (O_3) is one of the most important oxidants in the atmosphere alongside free radicals that include OH, HO_2, NO_3, and the halogens Cl, BrO, and I [8]. In the troposphere, the photochemical production of ozone is linked to the nitrogen oxides (NOx, $x = 1$ and 2) and HOx ($x = 1$ and 2) cycles [8, 9] as shown in Figure 4.3.

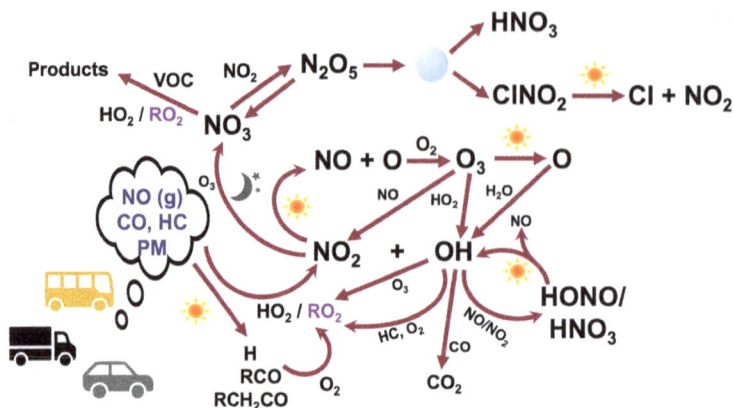

Figure 4.3: Sources and sinks of the ground level ozone in urban environments, impacted by traffic per reactions listed in references [10, 11].

At $\lambda < 424$ nm, the photolysis of nitrogen dioxide (NO_2) produces O atom, which quickly forms O_3:

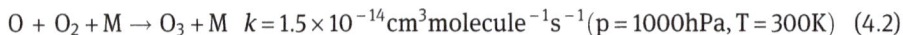

$$NO_2 + h\nu \rightarrow NO + O \quad j = 5 \times 10^{-3} s^{-1} \tag{4.1}$$

$$O + O_2 + M \rightarrow O_3 + M \quad k = 1.5 \times 10^{-14} cm^3 molecule^{-1} s^{-1} (p = 1000 hPa, T = 300K) \tag{4.2}$$

As a secondary pollutant, the ground levels of ozone can exceed 80 ppb in some locations around the world [12]. Tropospheric ozone is detrimental to human health and has been regulated globally since the nineties [12]. The World Health Organization (WHO) ozone air quality standard is 50 ppb, based on the daily maximum 8 h average level. In their analysis of the global ozone pollution, Fleming et al. [12] found that while the ozone levels across much of North America and Europe dropped significantly between 2000 and 2014, people in many US states, Europe, and China experienced ozone levels above 70 ppb for more than 15 days a year. Figure 4.3 shows the major gas phase sinks of ozone, which include the photolysis and reactions with NO and HOx. In their reviews, Jacob [11] and Usher et al. [13] provided detailed analyses of heterogeneous ozone sinks due to reactions with the aerosol particles and cloud droplets. In the following sections, selected examples are provided from more recent studies.

4.2.2.2 Heterogeneous reactions of ozone with organics

Ozone reactions with organic surfaces in atmospheric aerosol particles have been extensively studied using a variety of techniques that enabled quantification of the reactive uptake coefficient (γ) under dry and humid conditions, and direct monitoring of changes to surface functional groups to elucidate the reaction mechanisms. The γ

values refer to the probability that a gas phase molecule reacts with the surface, upon collision. Chapleski Jr. et al. [14] reviewed the literature on laboratory work using model systems. Table 4.2 lists the organic surfaces used and the major findings from these studies. Monitoring the ozone uptake reactions in situ using ATR-FTIR spectroscopy resulted in y values in the range from 1×10^{-5} to 5.1×10^{-4}, which was attributed to differences in the reaction conditions, surface properties, and the infrared band analyzed to extract kinetic information. Also, the y values were found to be higher by an order of magnitude on liquid ($\sim 1 \times 10^{-3}$), compared to frozen liquid samples ($\sim 1 \times 10^{-4}$), highlighting the effect of temperature and the participation of the underlying organic layers in the liquid samples in the reaction. Organics that form a monolayer, such as 1-octene, behaved like a frozen liquid sample.

The availability of $C = C$ double bond at the interface facilitates reactions with ozone, which were found to be invariant to surface film compression. Figure 4.4 provides a general scheme of the different reaction pathways that lead to the formation of different oxidation products, some of which are hygroscopic [15]. Water plays a role in product distribution because its amount – that differs at the interface compared to the bulk – affects the stability of some of the products through proton transfer. The role of RH in the heterogeneous ozonolysis reactions with organics varies with the phase of the organics (liquid versus solid) and the structure of the organic compounds [16]. Increasing RH was reported to have a retarding effect on the ozonolysis of organic particles, such as linoleic acid, suggesting a competition between O_3 and water molecules for the same adsorption sites [17]. On the other hand, increasing RH, and hence surface water coverage due to hydrogen bonding with catechol's OH groups, enhanced the ozonolysis rate of catechol at the air-solid interface. These results were explained by the ability of the surface water to facilitate an attack on the vicinal C1-C2 bond, resulting in the formation of muconic acid [18]. Criegee intermediates favor reactions with water than producing free radicals, which leads to the formation of smaller acids, resulting in more significant changes to the hygroscopicity of the organic particles [15, 17].

Ozonolysis reactions of the organics at the air-water interface have also been studied to mimic reactivity of the organic-coated aerosols. Woden et al. [19] used oleic acid monolayers, and varied the temperature and salinity of the water film to investigate their effect on the reaction with ozone. The results showed the formation of stable non-oxidizable organic products that persist at the air-water interface, particularly at near-zero degrees Celsius, with and without NaCl in the aqueous phase. The products identified in the study are nonanoic acid, azelaic acid, or 9-oxononanoic acid, and their ratios were affected by the composition of the aqueous phase.

Table 4.2: Heterogeneous reactions of ozone with the organic surfaces as summarized by Chapleski Jr. et al. [14] and the references therein.

Organic surface	Major findings
Polyaromatic hydrocarbon (PAH) Lab-generated soot Polycyclic aromatics Biogenic organics Biomass burning products Fungicides Functionalities at well-characterized aerosol surfaces	– Fast initial consumption of O_3, resulting in irreversible changes to surface functional groups, rendering the surface unreactive to further ozone exposure. – Two molecules of O_3 consumed for every molecule of CO_2 produced – $\gamma = 2 \times 10^{-7}$ on fresh soot, and 10^{-7} on 'aged' soot. – Amorphous carbon and disordered graphitic sites, responsible for initiating the reactions, which form surface products, that include surface-bound ketone, lactone, and anhydride groups.
Oleic acid NaCl/oleic acid particles Oleic acid/water droplet	– The $C = C$ bond location is at a non-terminal position, hindering the reaction with O_3 and decreasing the reactive γ values, compared to a monolayer with a terminal $C = C$ bond. – Reactive γ, in the rage of $10^{-5} – 10^{-3}$. Substrates with non-uniform coverage of oleic acid showed γ in the order of 10^{-5}. In general, γ values are highest on fresh surfaces and decrease with passivated surfaces. – Variability in γ is attributed to diffusion of O_3 into the bulk, secondary reaction pathways, beyond ozonolysis (experimental technique used), irrespective of whether γ values were calculated using changes in O_3 levels or in oleic acid concentrations/properties.
α-terpineol/glass PAH/glass Anthracene/glass	– Langmuir–Hinshelwood mechanism, whereby the impinging ozone molecule thermally accommodates on the surface before reacting.

Table 4.2 (continued)

Organic surface	Major findings
Air/naphthalene/water droplet	– Ozone and naphthalene react at the air-water interface at a rate 15x greater than the homogeneous gas phase reaction. – The rate of heterogeneous reaction is dependent on the size of the droplet. – Oxidation products with higher water solubility readily diffuse into the bulk → greater mass transfer of naphthalene from the gas phase to the droplet surface.
Air/saturated or terminally unsaturated phospholipid molecules/water	– No reaction with the saturated lipid molecules. – Reaction with the unsaturated compound (containing surface C = C double bond) yielded an aldehydic product.
Air/anthracene/water droplet	– Langmuir–Hinshelwood mechanism, where ozone first adsorbs to the air–aqueous interface and then reacts with the already adsorbed anthracene. – γ values ranged from 2×10^{-8} to 3×10^{-7}. – The presence of organic acids in the aqueous phase reduced the ozone reaction with the organic film. – The presence of alcohols enhanced the overall reaction rate. – Water, complex formation of organic acids with water or sites on anthracene, and the surface residence time of O_3 at the interfaces with and without organics, affect the reaction probability.
Aqueous microdroplets of uric acid, ascorbic acid, sulfonic acid, phenol, α-tocopherol, cysteine, β-caryophyllene	– Highlighted the differences in the ozonolysis reaction products between air-water shell and bulk solution because of changes in water density and the rate of vibrational relaxation, which allows for activation of difference processes.
Self-assembled monolayers with various terminations: alkene, methyl, hydroxyl, perfluoro	– Provided insights into the O_3-surface collision dynamics that include energy transfer, thermal accommodation coefficient, and reaction probability.

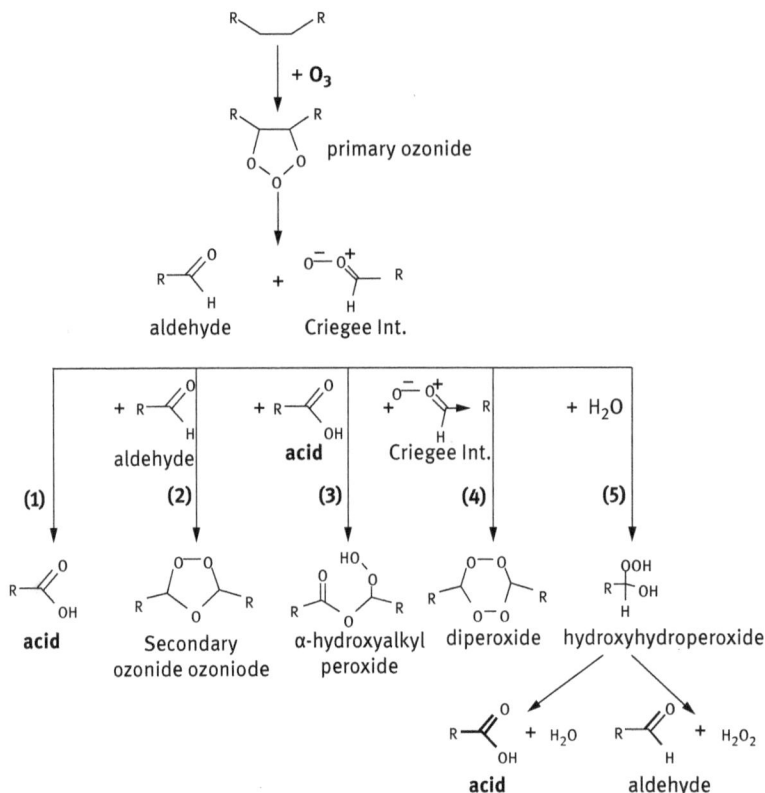

Figure 4.4: Reaction mechanism for the ozonolysis of an unsaturated double bond. Formation pathways of acids, aldehydes, secondary ozonides, α-hydroxyalkyl peroxides, and diperoxides in the presence and absence of water. Figure and caption were reproduced from reference [15] under the Creative Commons Attribution 3.0 License (CC-BY). © The Author(s), 2008.

4.2.2.3 Heterogeneous reactions of ozone with fresh and aged mineral dust

Field studies aimed at measuring the tropospheric ozone and particulate matter report anticorrelations between ozone mixing ratios and particles mass concentrations [13]. Modeling studies showed lower levels of ozone among other pollutants that include SO_2 and NO_x, in areas with high dust. For example, using γ values in the range $5.0 \times 10^{-5} \sim 2.0 \times 10^{-3}$, Dong et al. [20] simulated the impact of dust chemistry on ozone levels over eastern China as shown in Figures 4.5a and b after including the lower and upper γ values in the Community Multiscale Air Quality (CMAQ) modeling system, respectively. The concentration of O_3 was reduced by 3–6 ppbv (2–10%) and 5–11 ppbv (4–20%) from the lower and upper limits of γ, respectively.

These modeling results motivated laboratory studies on the uptake of ozone on authentic mineral dust, individual mineral, and metal oxide components, which were reviewed by Usher et al. [13] and Wang et al. [21]. The reactive sites on these

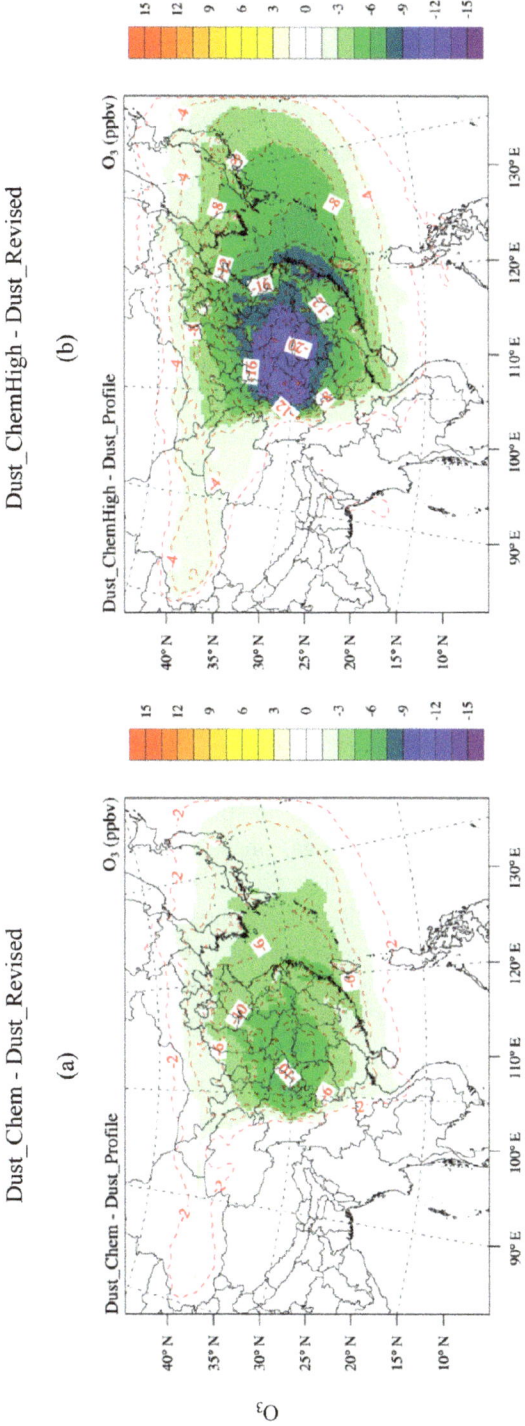

Figure 4.5: Five-year averages for March and April, from 2006 to 2010, of dust heterogeneous chemistry, and the impacts with lower (left column) and upper (right column) uptake coefficients for O_3 species. Color contours represent the absolute concentration changes and the dashed contour lines with numbers indicate the percentage changes. Dust_Chem: Same as Dust_Profile, but with implemented dust chemistry with a lower limit of uptake coefficient, Dust_ChemHigh: Same as Dust_Chem, but with an upper limit of uptake coefficients, Dust_Revised: Revised initial friction velocity threshold constant in the dust plume rise scheme, and Dust_Profile: Same as Dust_Revised, but with implemented source-dependent speciation profile. Figure and caption were modified from reference [20] under the Creative Commons Attribution 3.0 License. © The Author(s), 2016.

'fresh' materials are different from aged (i.e., previously reacted) ones, which would have a coating of salt (sulfate, nitrate) or organics. Lab experiments explored the Lewis acidity and basicity of ozone, upon exposure to unreacted metal oxides that were terminated with hydroxyl groups. Ozone either formed hydrogen bonds with hydroxyl groups through its terminal oxygen, as in the case of SiO_2, or underwent ligand exchange, as in the case of TiO_2. Reactions with basic metal oxides, such as CaO and MgO, formed weakly adsorbed ozone, whereas on alumina, the stronger Lewis acid sites dissociated ozone, upon adsorption. Experiments confirmed the catalytic nature of ozone destruction on metal oxides and dust particles [21], where a steady-state uptake was observed, with no indication of the ozone uptake rate approaching zero. This observation prompted the differentiation between the initial and steady-state γ values where the latter were found to be lower. For example, on α-Fe_2O_3 and α-Al_2O_3, the steady-state values were 2.2×10^{-5} and 7.6×10^{-5}, respectively, compared to the initial γ values of $1.8(0.7) \times 10^{-4}$ and $1.2(0.4) \times 10^{-5}$, respectively, at room temperature [13]. Nicolas et al. [22] examined the role of light on ozone uptake and reported a 10x increase in the γ values of ozone decomposition under irradiation versus dark conditions on SiO_2 mixed with trace amounts of TiO_2, from 3×10^{-6} to 3×10^{-5}. Also, increasing the RH values increased the photochemical uptake of ozone up to 20%, and above 30% RH, a reduction in photochemical ozone loss was observed. This behavior was explained by a competitive dual adsorption of water and ozone, where OH radical formation was facilitated at the mineral surface, and hence the ozone conversion, followed by a RH regime where dual adsorption inhibited the reaction [22].

Laboratory studies also investigated the ozone uptake on mineral dust proxies, with various degrees of inorganic and organic coatings, to simulate aged mineral dust. Using α-Al_2O_3 particles, previously reacted with gas phase nitric acid forming a nitrate coating, the ozone uptake was found to be lower by 70%, compared to the unreacted α-Al_2O_3 surface, judging by the reduction in γ values to $3.4(0.6) \times 10^{-5}$ from $1.2(0.4) \times 10^{-4}$. This reduction in reactivity was explained by the presence of fully oxidized surface species that blocked the reactive sites on alumina. The opposite trend was observed following the reaction with SO_2, forming surface sulfite (SO_3^{2-}), where a 30% increase in γ values was reported. Sulfite surface species were available for oxidation to sulfate by ozone, hence increasing particle reactivity. Similar observations were found in the case of organic coatings, terminated with reactive and oxidizable C = C functional groups on SiO_2 [13] and α-Al_2O_3 particles [23], as opposed to unreactive alkane groups. As reviewed by Tang et al. [24], studies to date show that the reactivity of mineral dust and its proxies with ozone had either no effect or a complex response of ice nucleation, compared to unreacted materials. Using kaolinite as a proxy for mineral dust, Hinrichs and co-workers [25] studied ozone decomposition on pure and monoterpene-processed particles as a function of RH. Monoterpenes limonene and α-pinene were chosen as they are abundant biogenic volatile organic compounds (VOCs) that contain C = C in their structure. Figure 4.6 shows a comparison between the average γ values for ozone uptake after correction for the BET surface area of the samples as a function of RH. This data show that at all humidities studied (10 – 50% RH), O_3 uptake was at least 1 order of magnitude higher for monoterpene-processed kaolinite that contain surface-adsorbed organics, compared to that of pure kaolinite. Competitive adsorption of water, with increasing RH, resulted in lower organic surface concentrations, and hence lower ozonolysis rates.

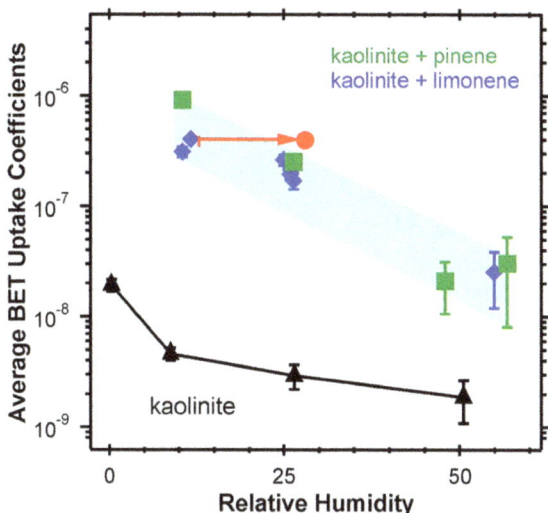

Figure 4.6: Average uptake coefficients assuming BET surface area, calculated between 4000 and 10,000 ppb-min O_3 dose, as a function of RH for pure kaolinite (black triangles), compared to kaolinite, pre-exposed to the α-pinene (green squares) and limonene (blue diamonds). Error bars show standard deviation range of uptake observed during the time frame. Red circle shows uptake for α-pinene, adsorbed at 12% RH but reacted with O_3 at 28% RH. Reproduced with permission from reference [25], © American Chemical Society, 2018.

The effect of light on ozone uptake and the ozonolysis of the organics adsorbed on the metal oxide and dust samples were also investigated by a number of groups, as reviewed by George et al. [26]. For example, Ma et al. [27] reported a 1.5x increase in the heterogeneous reaction rate between ozone and anthracene adsorbed on TiO_2 in the presence of light, compared to dark conditions. This rate was measured by monitoring the decay of adsorbed anthracene. TiO_2 is a minor component in mineral dust since it makes up about 0.68% of the abundant oxides in the continental crust [13]. However, it is a semiconductor and an efficient photocatalyst in the degradation of organics, and hence would contribute to the photoreactivity of mineral dust that generates ROS [28]. The ozonolysis rate of anthracene, adsorbed on Asian dust, was reported to be 4x faster in light, compared to dark conditions [27]. This study did not report the mineralogical composition of the Asian dust in the enhancement of the decay of surface anthracene, with irradiation. Nicolas et al. [22] proposed a mechanism for ozone photodecomposition based on the photocatalytic reactivity of

TiO$_2$, as shown in the following equations, which is likely at play when the organics are present on the surface of TiO$_2$:

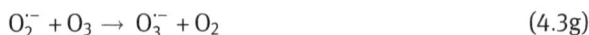

$$TiO_2 + h\nu \rightarrow TiO_2(e^- + h^+) \tag{4.3a}$$

$$O_3 + e^- \rightarrow O_3^{\cdot-} \tag{4.3b}$$

$$O_3^{\cdot-} + H^+ \rightarrow HO_3^\cdot \tag{4.3c}$$

$$HO_3^\cdot \rightarrow O_2 + HO^\cdot \tag{4.3d}$$

$$HO^\cdot + O_3 \rightarrow O_2 + HO_2^\cdot \tag{4.3e}$$

$$O_2 + e^- \rightarrow O_2^{\cdot-} \tag{4.3f}$$

$$O_2^{\cdot-} + O_3 \rightarrow O_3^{\cdot-} + O_2 \tag{4.3g}$$

The quantification of the γ values of ozone on mineral dust and its proxies has implication on the ozone atmospheric lifetime, τ. Lasne et al. [29] calculated τ from the available experimental data in the literature that reported the initial and steady-state γ values at low and atmospheric pressures on metal oxide and dust samples. They reported that the initial γ values from low pressure measurements are in the 10^{-6}–10^{-4} range, whereas the steady-state γ values are in the 10^{-10}–10^{-6} range. These ranges of γ values translate to $\tau \sim 10^1$–10^3 and 10^3–10^7 days, respectively. In order for heterogeneous ozone loss to compete with the gas phase pathways and reproduce the observations, the authors [29] highlight that γ values $> 10^{-5}$ would be significant for inclusion in models. These values were obtained in low pressure lab experiments of fresh dust and metal oxides. Since the γ values of aged dust with unsaturated organic coatings increase the ozone uptake, as mentioned above, the heterogeneous chemistry of these surfaces would be important for inclusion in models.

4.2.2.4 Heterogeneous reactions of ozone with halides

Halogens (X = Cl, Br, I) play an important role in the tropospheric depletion of ozone [10]. Simpson et al. [30] reviewed the sources, cycling, and impacts on halogen chemistry in the lower atmosphere. Figure 4.7 shows the interplay between the gas phase and the interfacial chemistry in halogen activation. The two multiphase reactions highlighted in this figure are the uptake of HX to the condensed phase, where it could deprotonate, forming X$^-$, and the interfacial reaction between HOX and X$^-$, forming X$_2$ and OH$^-$(aq). The enhancement of the heavier (more polarizable) halides, namely bromide and iodide, near the air-water interface (relevant to marine aerosol), relative to the bulk, has been reported using theoretical methods [31, 32] and experiments [33]. Upon freezing, these halides get excluded from the bulk to concentrate at the interface of the different regions in ice/snow, as highlighted in the review by Bartels-Rausch et al. [34]. The freezing process of solutions, containing halides and nitrogen oxides that mimic the inorganic content in sea spray, was also reported to produce tri-halide ions, leading to the release of interhalogen gases [35].

Heterogeneous reactions with the gas phase ozone could also take place, as reported by Clifford and Donaldson [36] in the case of aqueous bromide. They measured an increase in pH at the air-aqueous interface due to the following reactions that produced hydroxide ions in solution, $OH^-(aq)$:

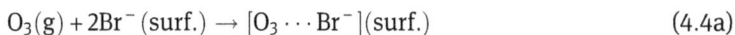

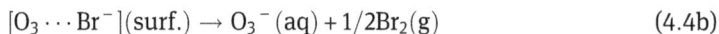

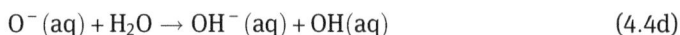

$$O_3(g) + 2Br^-(surf.) \rightarrow [O_3 \cdots Br^-](surf.) \tag{4.4a}$$

$$[O_3 \cdots Br^-](surf.) \rightarrow O_3^-(aq) + 1/2Br_2(g) \tag{4.4b}$$

$$O_3^-(aq) \rightarrow O_2(g) + O^-(aq) \tag{4.4c}$$

$$O^-(aq) + H_2O \rightarrow OH^-(aq) + OH(aq) \tag{4.4d}$$

$$\text{Overall: } O_3(g) + Br^-(surf.) + H_2O \rightarrow 1/2Br_2(g) + O_2(g) + OH(aq) + OH^-(aq) \tag{4.4e}$$

Oldridge and Abbatt [37] reported the formation of the gas phase bromine from the interaction of ozone with frozen and liquid mixed NaCl/NaBr solutions, and were able to quantify the contribution of surface chemistry from the bulk phase reactions. In particular, the dependence of γ values on ozone concentration indicate that bulk-phase kinetics dominate at high ozone concentrations, whereas surface phase kinetics is of greater relative importance at low ozone, with the fastest kinetics in acidic media. The authors also observed qualitatively similar dependencies of the kinetics on pH, bromide concentration, and ozone concentration when the samples were either aqueous or frozen solutions. These observations led the authors to believe that the chemistry – whether at the surface or in the bulk – occurs with brine formed when freezing the salt solution.

Wren et al. [38] investigated the heterogeneous reaction of the gas-phase ozone with bromide and iodide ions at the surface of frozen salt solutions. At an ozone concentration of $\sim 8 \times 10^{15}$ molecule cm^{-3}, and sea water concentrations of bromide and iodide, they reported ozone γ values on the ice surfaces of frozen solutions containing bromide to be $(1.3 \pm 0.5) \times 10^{-8}$ and iodide to be $(1.6 \pm 0.5) \times 10^{-9}$. These uptake coefficients were estimated to be greater than a factor of ten at atmospheric ozone concentrations, and $\sim 60 \times$ faster than at the surface of an aqueous solution at the same concentration.

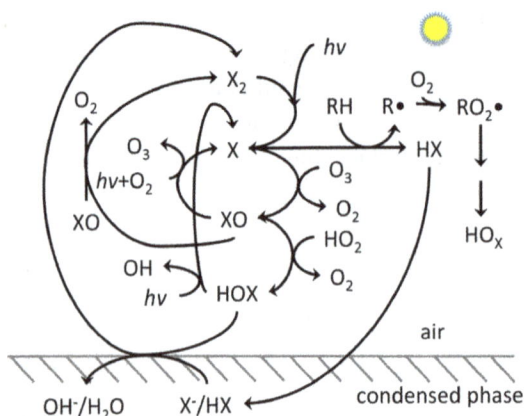

Figure 4.7: Simplified reaction diagram for halogen atoms, represented as "X" in this diagram, and key chemical reaction pathways. Note that many species on this diagram are radicals, but for simplicity, only organic radicals and organic peroxy radicals, denoted by R˙ and RO_2˙, are explicitly shown with an unpaired electron. The figure and caption were reproduced from reference [30] under the AuthorChoice License, © American Chemical Society, 2015.

4.2.3 Hydroxyl radicals (˙OH)

4.2.3.1 Sources and sinks of ˙OH in the gas phase

The hydroxyl (˙OH) radical is the most important oxidizing species during daytime and is therefore known as the '*detergent of the atmosphere*' [8, 10, 11, 39, 40]. Figures 4.3 and 4.7 show the daytime sources and sinks of ˙OH radicals in the gas phase. In the early morning hours of a polluted urban atmosphere, the photolysis of HONO and HCHO to form ˙OH radicals was reported (see for example, reference [41]). Stone et al. [42] reviewed the literature on the field measurements and model comparisons of the tropospheric ˙OH and $HO_2^{\cdot}$ radicals. Because of its high reactivity with a wide range of compounds, its concentration is extremely low, and ranges from 10^5 to 10^7 molecules cm^{-3}. Sinks for ˙OH radicals produced in the atmosphere are shown in Figure 4.3 and they include reactions with CO, hydrocarbons including methane, ozone and NO_x ($x = 1$ and 2). On a global scale, 40% of the ˙OH radicals react with CO, 30% with organics, 15% with methane, and 15% with O_3, $HO_2^{\cdot}$, and H_2. Reactions with organics proceed first by hydrogen abstraction in the case of alkanes and alkenes, followed by ˙OH addition in the case of alkenes and CO, as shown in the following equations:

$$1) \text{ Alkanes } (RH) + {\cdot}OH \rightarrow R{\cdot} + H_2O \tag{4.5a}$$

$$R{\cdot} + O_2 \rightarrow ROO{\cdot} \tag{4.5b}$$

$$2) \text{ Alkenes } (RCH{=}CH_2) + {\cdot}OH \rightarrow RCH - CH_2OH \tag{4.5c}$$

$$RCH-CH_2OH + O_2 \rightarrow RCHOO^{\cdot}-CH_2OH \qquad (4.5d)$$

$$3)\,CO + {\cdot}OH \rightarrow [HOCO] \rightarrow H^{\cdot} + CO_2 \qquad (4.5e)$$

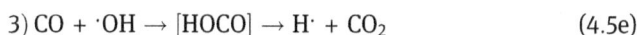

In their review articles, Gligorovski et al. [39] and Atkinson and Arey [43] provide additional details on the above reactions, including mechanisms and rate constants. George and Abbatt [44] reviewed the heterogeneous oxidation of organic and inorganic components in the atmospheric aerosol particles by gas phase radicals, including the ${\cdot}$OH radicals. The following sections describe case studies that highlight the heterogeneous loss of ${\cdot}$OH radicals from reactions with surfaces containing organics and halides.

4.2.3.2 Heterogeneous loss reactions of ${\cdot}$OH with organics

Reactions of ${\cdot}$OH radicals with atmospheric organic aerosols received much attention because these oxidation reactions would transform the surface functional groups and hence impact the physicochemical properties of atmospheric particles. Chapleski Jr. et al. [14] reviewed the literature on the heterogeneous reactions of ${\cdot}$OH radicals with organic surfaces that include squalane ($C_{30}H_{62}$, liquid at 25 °C), squalene seed coated with viscous secondary organic aerosol (SOA) material from α-pinene ozonolysis octacosane ($C_{28}H_{58}$, solid at 25 °C), and other organic compounds that are of atmospheric relevance in the form of aerosols, thin film coatings on glass flow tubes or optical windows for infrared spectroscopy measurements. Their synthesis of the literature in that review [14] highlighted the following mechanistic details: (1) lab measured y values have been determined from either measuring the loss of ${\cdot}$OH radicals or particle loss; are generally high and range between 0.2–2, where higher than unity values indicate secondary reactions taking place, (2) oxygen concentration, molecular structure (that affect functional group accessibility), and phase of the organics affect reactivity with ${\cdot}$OH radicals; with liquids reacting faster than solids), and terminus functional groups having a higher likelihood of functionalization or carbon loss (depending on oxidation lifetimes), and (3) the mechanism is consistent with the Langmuir-Hinshelwood mechanism as ${\cdot}$OH radicals prefer interfacial layers over the bulk. For example, in a report by Sato and co-workers [45], the aging of limonene SOA by OH radicals resulted in the increase of SOA mass concentration by 25% along with the degradation of gaseous unsaturated products – limononic acid, limononaldehyde, limonaketone, and limonalic acid, and particulate products – limonic acid and hydroxylimononic acid.

Moreover, Heath and Valsaraj [46] measured rate constants for the reaction of ${\cdot}$OH radicals with aqueous films enriched with benzene at the interface to simulate fog droplets. Varying film thickness allowed for calculating the bulk and interfacial contributions to the overall observed rates. Interfacial rate constants exceeded the bulk phase rate constants by 4 to 5 orders of magnitude. Reducing the film thickness greatly increased the reaction rate, while increasing the ionic strength to

0.025 M and adjusting pH in the range of 3.20 to 7.36 had no measurable effect on the overall rate constants.

To examine the effects of the liquid-liquid phase separation (and hence morphologies) and RH on the heterogeneous ˙OH oxidation of inorganic–organic aerosols, Lam et al. [47] used aqueous droplets containing 3-methylglutaric acid (3-MGA) and ammonium sulfate (AS) particles with an Organic-to-InoRganic Dry mass ratio of 1 (OIRD1) as models using aerosol mass spectrometry. Figure 4.8a shows the effect of RH on the signal decay, as a function of ˙OH exposure. These data were used to extract values of the ˙OH second-order heterogeneous reaction rate constant (k), as shown in Figure 4.8b. This model aerosol system has phase separation relative humidity (SRH) between 72–75%, as measured by two techniques. Phase separation occured below SRH and the aerosol had a well-mixed phase above SRH. The data in Figure 4.8 show that with decreasing RH, from 88% to 55%, the k value increased from $1.01 \pm 0.02 \times 10^{-12}$ to $1.73 \pm 0.02 \times 10^{-12}$ cm^3 molec.$^{-1}$ s^{-1}, suggesting that the heterogeneous ˙OH reactivity of 3-MGA in the phase-separated droplets (RH < SRH) is slightly higher than that of aqueous droplets with a single liquid phase (RH > SRH). Images of the 3-MGA/AS droplets as a function of RH showed that 3-MGA had an inhomogeneous distribution in the phase-separated particles that adopted the core-shell and the partially engulfed morphologies with higher concentrations present at/near the droplet surface, which increases the collision frequency of ˙OH radicals with the organics.

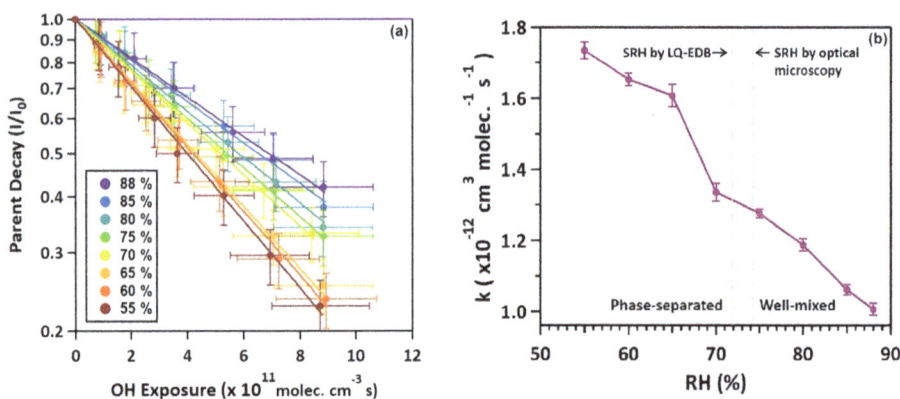

Figure 4.8: (a) Normalized parent decay of 3-MGA/AS aerosols (OIRD1) as a function of RH; (b) the effective second-order heterogeneous OH reaction rate constant (k) plotted against RH for the heterogeneous OH oxidation of 3-MGA/AS aerosols (OIRD1). SRH = separation relative humidity, LQ-EBD = linear-quadrupole electrodynamic balance. Figure and caption were modified from reference [47] under the Creative Commons Attribution 4.0 License, © The Author(s), 2021.

4.2.3.3 Heterogeneous loss reactions of ˙OH with halides

George and Abbatt [44] reviewed the reactions of ˙OH radicals with inorganics that include halides and bisulfate. Halide oxidation takes place in these reactions, leading to the release of Cl_2 and Br_2 gases. The following reactions were proposed based on experimental data with NaCl and NaBr particles:

$$^\cdot OH(g) + Cl^-(surface) \rightarrow HO \cdot Cl^-(surface) \quad (4.6a)$$

$$HO \cdot Cl^-(surface) + HO \cdot Cl^-(surface) \rightarrow Cl_2(g) + 2OH^- \quad (4.6b)$$

$$^\cdot OH(g) + 2Br^-(surface) + 2H^+ \rightarrow Br_2 + 2H_2O \quad (4.7)$$

Sjostedt and Abbatt [48] quantified the gas phase product yields from the reaction of $^\cdot OH$ with NaCl/NaBr, with iodide present as an impurity in the form of frozen salt solutions, humidified desiccated salts, and desiccated salts under dry air. The ratios of the halides were adjusted to those found in seawater, that is, $[Cl^-]/[Br^-] = 680:1$ and $[Cl^-]/[I^-] = 109,000:1$. In these systems, the detectable amounts of the oxidation products Br_2, BrCl, and IBr gases were reported mostly from acidified frozen solutions.

4.2.4 Reactive Nitrogen Species (RNS)

4.2.4.1 Sources and sinks of RNS in the gas phase
Figure 4.3 shows the daytime formation and the reactions of RNS species that include the radicals, NO and NO_2, and the nonradicals, HONO and HNO_3. Reactions of the NOy species (i.e., $NO + NO_2 + HNO_3$) in the troposphere are closely linked to the HOx cycle [10, 49] that leads to the photochemical ozone[9], hydroxyl radical [10, 49], and SOA formation [50]. Because HONO photolysis leads to $^\cdot OH$ formation, the homogeneous and the potential heterogeneous sources of HONO received special attention, as highlighted in the following text. During nighttime when ozone and hydroxyl radicals are consumed at higher rates than their formation, the nitrate radical, $NO_3^\cdot$, becomes the main oxidizer in the atmosphere [11, 51]. This radical forms from the oxidation of NO_2 by ozone, and its major sinks are rapid photolysis ($\lambda < 590$ nm), forming NO_2 and $O(^3P)$, and the reactions with VOCs and NO_2 that eventually lead to the formation of peroxyacetyl nitrate (PAN), N_2O_5, and nitric acid, respectively [43, 52]. The following sections describe the heterogeneous reactions of RNS, and their role in affecting the budget of gas phase RNS and changing the physicochemical properties of the atmospheric aerosol particles.

4.2.4.2 Heterogeneous reaction of RNS with organics
The heterogeneous reactions of the radical and non-radical RNS with organic surfaces have been the subject of several reviews. More specifically, the heterogeneous chemistry of NO_2, HNO_3, N_2O_5, and $NO_3^\cdot$ radical with organics at the surfaces are described below. These studies were partially motivated by observations and modeling studies,

suggesting missing sources of RNS, such as HONO, in a number of locations around the world [53].

Ma et al. [54] reviewed the literature on the dark and photochemical reactions of NO_2 with organic aerosol. More specifically, heterogeneous reactions of NO_2 with soot received much attention as a potential reaction that leads to nighttime formation of HONO. Soot samples were generated by different methods, which led to a wide range of γ values of NO_2 in the dark ($10^{-1} - 10^{-8}$) and HONO yields (few percent to 100%). Under atmospheric conditions, soot surfaces were found to deactivate in the dark because of coatings and irreversible functionalization with nitrite groups, making their surface chemistry with the gas phase NO_2 a minor source of HONO in the troposphere (see for example, references [55–57]).

Under simulated solar irradiation, photoenhanced uptake of NO_2 and photoproduction of HONO on soot [58] and a wide range of organic surfaces was reported [54, 59, 60]. The values of γ for NO_2 under UV radiation ranged from 2×10^{-5} on soil humic acid to 5×10^{-7} on soot in the RH range of 20–56% [59]. The photo yield of HONO per reacted NO_2 ranged from 57–97% for different internally- and externally-mixed organic films, compared with 0–11% in the dark, as summarized by Ma et al. [54]. The possible reaction mechanisms proposed in the previous studies centered on the formation of electronically photoactivated surface species, such as the aromatic and carbonyl functional groups, which undergo electron transfer reactions that form reductive and oxidative centers on the organic surface, eventually reducing NO_2 and forming HONO from the reaction of NO_2 with the reductive centers. Through a series of careful experiments, Raff and co-workers [61] characterized the surface functional groups and quantified the NO_2 uptake and HONO formation in the dark, and under irradiation in soil surrogates composed of humic acid and clay minerals. Their results suggested that hydroquinones and benzoquinones contribute to the thermal and photochemical HONO formation. Figure 4.9 shows the quinone redox chemistry and the potential intermediates responsible for NO_2 to HONO conversion by organic matter, highlighting the role of redox active sites [61]. The group also observed the abundance of C-O moieties in humic acid correlated with the conversion of NO_2 to HONO, whereas the photochemical reactivity correlated with the abundance of C = O moieties.

Nitric acid (HNO_3) forms in the gas phase from the oxidation of NO_2 by the hydroxyl radicals (Figure 4.2). Nitric acid has a relatively high Henry's law constant ($0.88–26 \times 10^3$ mol m^{-3} Pa^{-1} at 298 K [62]), which is 3–5 orders of magnitude higher than HONO. Condensation reactions with ammonia were found to lead to a rapid growth of new atmospheric particles [63]. The heterogeneous reactivity of HNO_3 with the inorganic components in atmospheric aerosol particles received much attention and are detailed in the sections below. Ault et al. [64] collected nascent sea spray aerosol (SSA), which contain salt, organic, and biological material, and monitored their reaction with gas phase nitric acid. Figure 4.10 shows a summary of the reactivity results of individual SSA particles. Three particle types were described: (a, top) particles with high reactivity – evident by the relatively high nitrate signal

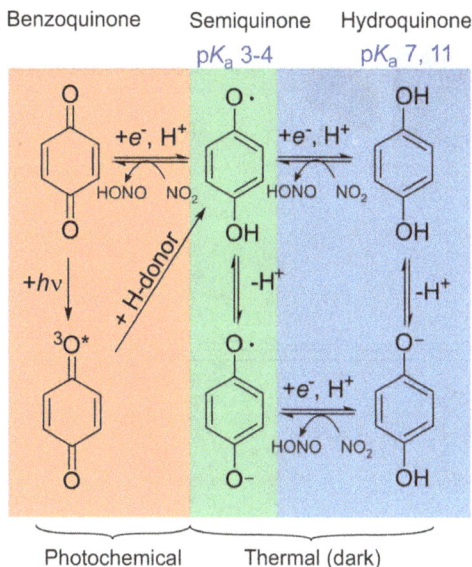

Figure 4.9: A schematic showing the quinone redox chemistry and the potential intermediates responsible for NO₂-to-HONO conversion by organic matter. The color shading indicates the oxidation state (from left to right): oxidized, partially reduced, reduced species, respectively. Figure reproduced with permission from reference [61], © American Chemical Society, 2017.

in the Raman spectrum – with minimum organic content, (b, middle) particles with low reactivity associated with high organic content, and (c, bottom) particles with intermediate reactivity, with peaks associated with organic species and nitrate. Figure 4.11 shows the three main chloride-displacement reactions in SSA, containing lipopolysaccharides (LPS) as a biological component and also containing carboxylate and phosphate groups [65]. These results highlighted the need for using a range of γ values for HNO_3 on SSA in atmospheric chemistry and climate models.

As shown in Figure 4.3, the reaction of NO_2 with NO_3 leads to the formation of dinitrogen pentoxide (N_2O_5), which has a Henry's law constant of 2.1×10^{-2} mol m^{-3} Pa^{-1} at 298 K [62]. Field observations and the heterogeneous chemistry of N_2O_5 received much attention as a NOx reservoir and has been reviewed by Chang et al. [66]. For example, the hydrolysis of N_2O_5 forms HNO_3 according to the following reaction:

$$N_2O_5 + H_2O \rightarrow 2HNO_3 \tag{4.8}$$

N_2O_5 was reported to effectively activate particulate chloride at nighttime, forming $ClNO_2$, which rapidly photolyzes to NO_2 and Cl radicals according to the following reaction [30]:

$$N_2O_5 + Cl^- \rightarrow ClNO_2 + NO_3^- \tag{4.9}$$

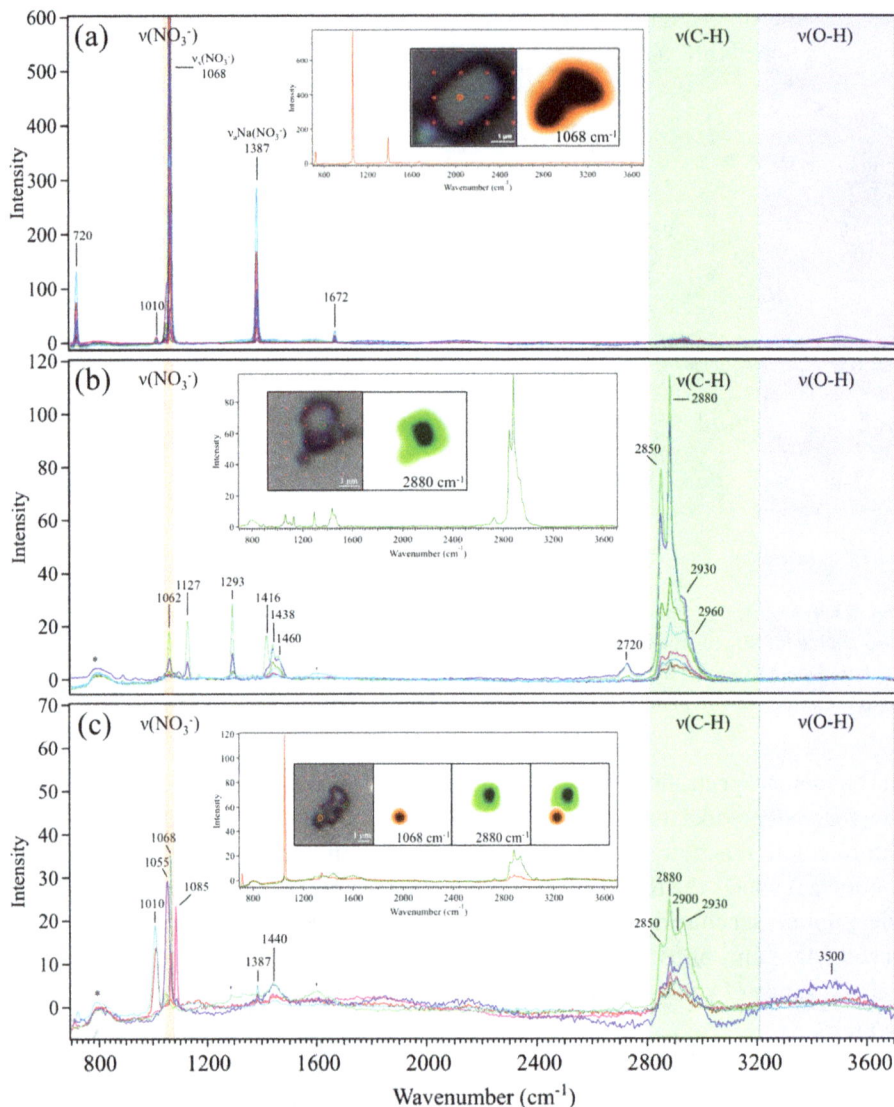

Figure 4.10: Raman spectra of individual SSA particles, analyzed with micro-Raman spectroscopy for several different particle types, along with optical images, individual spectra, and spectral maps in the insets. The larger colored circles on the optical images indicate the locations of the particles that the laser beam interrogates for each of the spectra shown in the inset. The smaller dots represent the grid that defines the points used for the spectral maps. These spectra of particles that are exposed to nitric acid show the different particle types. (a) Particles with high reactivity, as indicated by an intense signal for the symmetric stretch of the nitrate ion at 1068 cm^{-1}, along with other nitrate vibrations. These particles also show minimal intensity in the C–H stretching region, indicating small amounts of organic species present within these particles. (b) Particles with high intensity due to organic species, as shown by the large intensity in the C–H

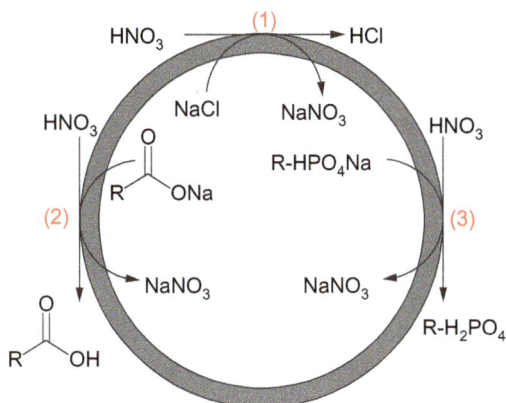

Figure 4.11: Nitric acid can react with different components of SSA as summarized above. (1) is the well-known chloride displacement reaction. Other reactive sites within the bioaerosols, in particular that of LPS, include carboxylate (2) and phosphate (3) groups. Figure and caption were reproduced with permission from reference [65], © American Chemical Society, 2016.

Because of the reactivity of chlorine atoms in the troposphere that shortens the lifetime of hydrocarbons, and because mercury catalyzes the formation/destruction of ground-level ozone, several groups quantified the yield of $ClNO_2$ from the above reactions in marine [67, 68], continental [69], and polluted regions [70, 71]. Moreover, laboratory investigations measured the uptake coefficients of N_2O_5 on ambient aerosol particles [66, 70–74], proxies for organic aerosols [75–78], organic coatings on SSA [79], and internally mixed inorganic and organic particles [80]. As summarized by McDuffie et al. [74], the $\gamma(N_2O_5)$ values on ambient aerosols collected during aircraft, ship, and ground campaigns are in the range from $10^{-3} - 10^{-1}$. Their analysis of field and lab work suggested that aerosol water is the dominant controlling factor of ambient $\gamma(N_2O_5)$. The effect of organic coatings on SSA at RH greater than 50% was found to not significantly limit the N_2O_5 reactive uptake coefficient in

---

Figure 4.10 (continued)
stretching region, indicating high levels of organics associated with the particle, but minimal signal associated with the nitrate ion. These particles have low reactivity. (Note: Although there is a peak at 1062 cm^{-1}, close to the NO_3^- region, it is most likely due to vibrations associated with the organic species within these particles, for example, C–C stretches, C–O stretches, and/or methyl deformations, associated with the organic species present, as its intensity correlates with other peaks due to organics from 1100 to 1500 cm^{-1}.) (c) Particles with both peaks associated with the organic species and a nitrate anion, indicating particles with intermediate reactivity. Spectral maps are shown for the $v_s(NO_3^-)$ mode (1068 cm^{-1}) and $v(C–H)$ mode at 2880 cm^{-1}. For these particles, the spectral maps along with the optical image show phase segregation between the nitrate and organic species within the particle. * designates the quartz substrate, and ' designates a small amount of graphitic carbon from sample damage by the laser. Figure and caption reproduced with permission from reference [64], © American Chemical Society, 2014.

the range 0.01–0.03 [79]. In the case of organics internally mixed with ammonium bisulfate [80], the effect of organics on $\gamma(N_2O_5)$ was less pronounced at RH > 50%. Higher $\gamma(N_2O_5)$ were observed using organics with O:C molar ratio > 0.56, which was attributed to the effect of liquid-liquid phase separation that most readily occur for organic compounds with O: C ratios of ≤ 0.7. Knopf et al. [75] estimated the atmospheric lifetime, τ, of biomass burning aerosol (BBA) to range from 1–112 min, which accounts for the degradation of one monolayer coverage of BBA surrogates due to the reaction with N_2O_5 and other gas phase RNS, for concentrations typical of polluted environments at 298 K.

Chapleski et al. [14] and Ng et al. [52] focused on the nitrate radical chemistry with liquid and solid organic substrates that include BBA, alkanes, alkenes, proxies for fatty acids with various degrees of saturation, and functionalized self-assembled monolayers. The studies focused on measuring the uptake coefficient, the effect of RH, the phase of the organics (liquid vs. solid), the reaction mechanism, the gas phase products and the changes to the surface functional groups. The reported values of γ on the single component organic surfaces differ by 1–3 orders of magnitude as shown in Figure 4.12, depending on the organic surface and the experimental conditions. Uptake on frozen samples either resulted in a decrease or no change in γ values, depending on the organic material. These results highlighted how the reaction mechanism affects factors that lead to surface site renewal through surface or bulk diffusion. Like the hydroxyl radical, the nitrate radical reacts with the organics

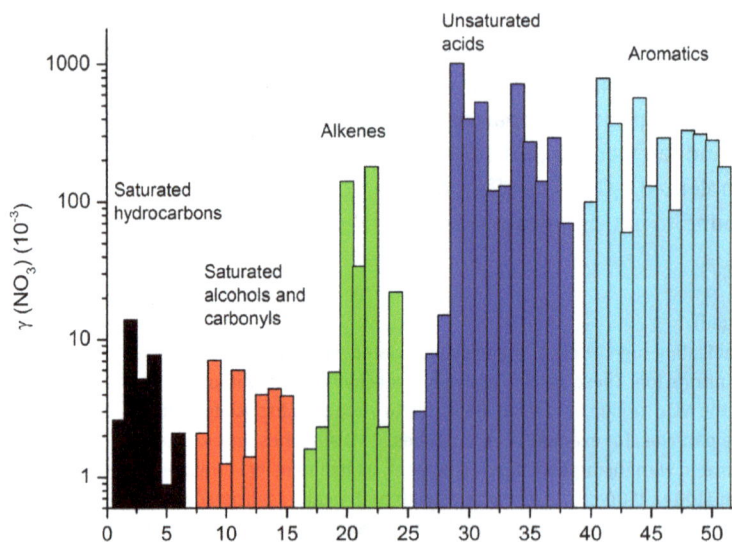

Figure 4.12: Uptake coefficients, $\gamma(NO_3)$, for the interaction of NO_3 with single-component organic surfaces. Details of the experiments and the references (corresponding to the x-axis numbers) are given in Table S1 in the Supplement of reference [52]. Figure and caption were modified from reference [52] under the Creative Commons Attribution 3.0 License (CC-BY). © The Author(s), 2017.

through either hydrogen abstraction or addition to a $C = C$ double bond, with NO_2 and O_2 being critical co-reactants. The kinetics of the second mechanism was found to be dominated by surface rather than bulk contributions, where a surface-bound radical is formed and is reactive toward gas phase species that scavenge radicals and, further, oxidize the surface.

4.2.4.3 Heterogeneous reactions of RNS with mineral dust

The uptake of some RNS on mineral dust, metal oxide, and clay was investigated in the lab to quantify the heterogeneous uptake coefficients as a function of RH and chemical composition, identify and quantify the conversion of RNS reactants to other forms of reactive RNS, and investigate the changes to the hygroscopic properties of the particles. The main goal of these studies was to assess if surface chemistry would provide additional sink/source pathways that competes with the known gas phase chemistry shown in Figure 4.3. Several reviews [13, 81, 82] synthesized the literature on these topics that covered the heterogeneous chemistry of NO_2, NH_3, HNO_3, NO_3, N_2O_5. The major findings from these reviews are as following. Metal oxides in mineral dust such as SiO_2, Al_2O_3, Fe_2O_3, and TiO_2 provide amphoteric sites for adsorption and surface reaction. Reaction products vary, depending on the amount of surface water and the type of metal oxide. For example, under dry conditions, HNO_3 adsorbs molecularly and reversibly on SiO_2, whereas it dissociates, upon adsorption, on Al_2O_3, Fe_2O_3, and TiO_2, forming oxide-coordinated nitrate surface species according to the following reaction:

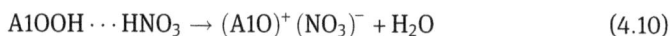

$$AlOOH \cdots HNO_3 \rightarrow (AlO)^+ (NO_3)^- + H_2O \qquad (4.10)$$

In the case of basic oxides, such as CaO and MgO, surface reactions with HNO_3 lead to the formation of solid nitrate salts, such as $Mg(NO_3)_2$ and $Ca(NO_3)_2$. Hydroxyl groups on these oxides also facilitate the formation of nitrate salts according to the following reactions:

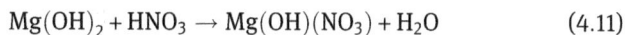

$$Mg(OH)_2 + HNO_3 \rightarrow Mg(OH)(NO_3) + H_2O \qquad (4.11)$$

$$Mg(OH)(NO_3) + HNO_3 \rightarrow Mg(NO_3)_2 + H_2O \qquad (4.12)$$

The extent of these reactions increase with increasing surface water because of site regeneration from the bulk. These reactions were found to increase the roughness of the metal oxide particles due to forming salts that have different morphologies than the underlying oxide. The experimentally determined initial uptake coefficient, γ_0, for HNO_3 on dry metal oxides ranged from 6.9×10^{-5} for α Fe_2O_3 to 1.6×10^{-2} for CaO. The γ_0 value measured for Gobi dust is in the middle, at 1.1×10^{-3}. These γ^* values take into account the surface area, porosity of the solid materials, and multiple collisions caused by the roughness of the individual particles. Surface deactivation occurred with time during the experiments, which lowers the γ_0 values. The carbonate content in mineral dust was found to be highly reactive with the gas phase HNO_3, particularly under humid conditions, according to the following reaction:

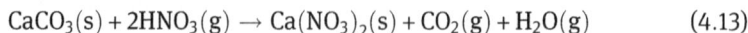

$$CaCO_3(s) + 2HNO_3(g) \rightarrow Ca(NO_3)_2(s) + CO_2(g) + H_2O(g) \tag{4.13}$$

Scanning electron microscopy images of single carbonate particles showed the transformation to aqueous droplets due to the deliquescence of the calcium nitrate product. As found with the basic metal oxides, this reaction was found to be not limited to the surface, making the total particle content of calcium carbonate available for reaction. Also, the reaction of HNO_3 with NH_3 on the surface of metal oxides led to the formation of NH_4NO_3 coatings that deliquesce with increasing RH, at values close to pure NH_4NO_3 particles. The presence of the metal oxide inclusions facilitated efflorescence of NH_4NO_3, with decreasing RH between 8–10%, which was not observed in pure particles due to the large kinetic barrier to crystallization in homogeneous nucleation. As summarized by Usher et al. [13], reactions of nitric acid with the alkaline content in mineral dust have implications on the optical and hygroscopic properties of the particles and on the HNO_3 to NO_x ratio in the gas phase.

In the case of NO_2, heterogeneous hydrolysis was reported on glass surfaces containing silicates, producing HONO and NO as the two major gas phase products. Other studies reported the formation of adsorbed HNO_3 on SiO_2 particles. Heterogeneous reaction of NO with adsorbed HNO_3 was found to occur faster than in the gas phase, with adsorbed water playing a crucial role according to the following equation:

$$NO(g) + 2HNO_3(ads) + 3H_2O(ads) \rightarrow NO_2(g) + N_2O_4(ads) + H_2O(l) \tag{4.14}$$

Adsorbed nitrito (NO_2^-) and then nitrate (NO_3^-) groups were detected with increasing reaction time of NO_2 with metal oxides in a stoichiometric manner that leads to surface saturation. Gas phase NO was produced after an induction period. The experimentally determined γ_0 for NO_2 on dry metal oxides ranged from 2.2×10^{-8} for γ–Al_2O_3 to 8.5×10^{-5} for α–Al_2O_3. The γ_0 value measured for Gobi dust is 4.4×10^{-5}.

The heterogeneous uptake coefficients of HONO, N_2O_5 and NO_3 radicals were also measured on mineral dust and metal oxides as summarized by Tang et al. [82]. Some of the studies quantified the effect of RH on the initial γ_0, which decreases with increasing RH for HONO. Higher γ_0 values by ~2 orders of magnitude were reported when illuminating TiO_2 and Al_2O_3, indicating a photoenhanced uptake of HONO. Results were mixed for N_2O_5, where increasing RH decreased γ_0 on illite and increased it on calcite, with no significant effect on the Saharan dust. Moreover, the heterogeneous uptake of the NO_3 radical was studied on Saharan dust, calcite, kaolinite, limestone, and Arizona test dust. The initial and steady state γ values were lower than those for N_2O_5, suggesting that the reactions with mineral dust are not important sinks for tropospheric NO_3 radicals except in regions with heavy dust loadings.

The above studies on model aerosol system motivated similar ones on soil particles. The role of surface acidity of metal oxides in HONO uptake was investigated by Donaldson et al. [83] They found minimum uptake coefficients near neutral soil pH, which is ~3 pH units higher than that observed for the minimum uptake into bulk water (pH 2–3). These results were interpreted in the context of the point of zero charge (pH_{PZC}) of amphoteric metal oxides in soil, particularly iron and aluminum (oxyhydr)oxides with pH_{PZC} in the 5–11 range. The interaction of these sites with HONO leads to highest uptake coefficients at soil pH above and below pH_{PZC}. Under basic conditions (pH > pH_{PZC}), solvated nitrite, bonded nitrate ($-ONO_2^-$), nitrito ($-ONO^-$) and nitro ($-NO_2^-$) groups formed upon HONO uptake, whereas under acidic conditions (pH < pH_{PZC}), nitroacidium ion (H_2ONO^+) formed. This study also showed that HONO is emitted in higher amounts under acidic soil pH containing nitrite due to the reaction of soil NO_2^- with protonated sites. This study highlighted an important role for the terrestrial-atmospheric cycling of nitrogen, with implications on HONO sources, air pollution, and the climate.

4.2.5 Heterogeneous reactions of volatile and semi volatile organics with mineral dust and its components

Field and laboratory studies reported heterogeneous reactions between biogenic and anthropogenic VOC with mineral dust, as summarized in earlier reviews [13, 82], to assess the importance of these loss pathways, relative to gas phase losses that include photolysis and reactions with ROS and RNS. These reviews highlighted the role of metal oxides and clays in providing sites for adsorption and reaction that coated the surface of the particles with organic functional groups. For example, aldol condensation reactions of aldehyde and ketone compounds, such as acetaldehyde, propionaldehyde, and acetone, occurred on amphoteric oxide particles, following adsorption. Adsorption of acids, such as acetic acid, and alcohols, such as methanol, led to deprotonation and formation of strong hydrogen bonds on the surface, as revealed from spectroscopic studies, leading to partial irreversibility and surface saturation with increase in surface coverage. Uptake coefficients, corrected for the BET surface area of the oxide particles, were in the range between 10^{-6} – 10^{-4}, depending on the type of the VOC and the metal oxide – the least reactive of which is SiO_2. Adsorption of alkanes and other classes of VOC on metal oxide and dust materials was also observed with a decreasing extent of uptake as RH increases from 10–90%. As reported by Starokozhev et al. [84], who investigated the uptake of VOC in multiphase systems, the partitioning is a complex process consisting of several stages. The interaction between the organic gas phase compounds and the condensed phases can often limit the partitioning and can be quantified by adsorption kinetics. Other factors that affect the interaction of VOC with water and solid interfaces, including mineral dust, and determine their partition behavior are functional groups and the hydrocarbon chain length. In a related study, Zeineddine et al. [85]

investigated the heterogeneous reactions of isoprene – the highest emitted biogenic volatile organic compound (BVOC) in the atmosphere – with natural Gobi dust, as a function of temperature and RH. Figure 4.13 shows the dependence of BET surface area-corrected γ_0 values on RH at 296 and 325 K. A nonlinear inverse relationship between the γ_0 values and RH was observed, and the adsorption process was dominated by reversible weakly bonded species. The atmospheric lifetime of isoprene based on this heterogeneous pathway was found to be in the order of hours to days, which is much longer than that calculated from the fast gas phase loss reactions with NO_3 and OH radicals (minutes to less than a day). Hence, the role of RH in the adsorption and surface reactivity is VOC is oxide/clay-dependent, and warrants careful and systematic investigations.

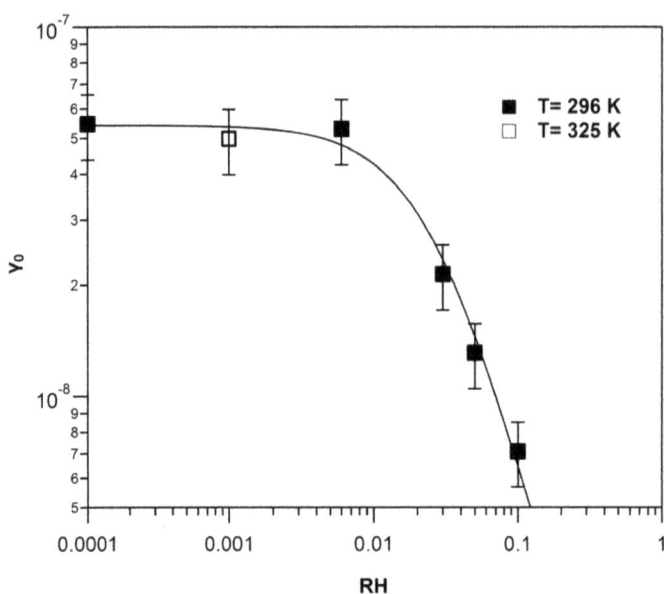

Figure 4.13: Initial uptake coefficients determined over the RH range of 0.01–10% as a function of the isoprene gas phase concentration, monitored by SIFT-MS. The error bars represent the overall estimated uncertainty on γ_0. The uptake coefficient was found to be inversely dependent upon RH, according to the empirical equation (solid line fitting the experimental points): $\gamma_0 = (2.7 \times 10^{-10})/(0.005 + RH^{1.44})$. Figure and caption were reproduced with permission from reference [85], © American Chemical Society, 2017.

Several theoritical studies were published after investigating the heterogeneous reactions of NO, NO_2, SO_2, and O_3 with mineral dust and metal oxide particles, following the adsorption of organics. Using density functional theory (DFT) calculations, Ji et al. [86] showed that the adsorption of HCHO on SiO_2 model cluster was more energetically favorable than NO_2 and that the presence of SiO_2 could accelerate the atmospheric reaction rate of formaldehyde with NO_2 to produce HONO, which decreased with decreasing temperature. In a related study by Wang et al. [87], DFT calculations were reported on the gaseous and heterogeneous ozonolysis reactions of low-molecular-weight (LMW) unsaturated ketones with an SiO_2 model cluster. The ketone compounds included 3-buten-2-one (BO32), 3-penten-2-one (PO32), 4-penten-2-one (PO42), and 4-methyl-4-penten-2-one (MPO442). Adsorption of LMW on SiO_2 proceeded via hydrogen bonding between the carbonyl oxygen and the hydrogens in the silanol groups. The reactivity of these compounds with ozone in the gas phase and on the surface of $Si(OH)_4$ decreased in this order: PO32 > MPO442 > PO42 > BO32, with aldehydes, diketones, and some organic peroxides as the main products. The heterogeneous reactions were found to be 1–5 times faster than those in gas phase. In general, the results suggested that SiO_2 is an efficient sink of LMW unsaturated ketones and the corresponding oxidation products, especially in desert areas and dusty environments.

The heterogeneous chemistry of iron (oxyhydr)oxides, such as hematite, goethite, and ferrihydrite, with organics have also been studied (for reviews, see references [88–92]). One example that highlights the role of RH in the adsorption of catechol vapor is shown in Figure 4.14. Because hematite is an insoluble metal oxide, catechol adsorption occurs via exchange with ligands on Fe sites as shown in the following equation:

$$2 \equiv FeOH + Catechol \rightarrow [\equiv Fe\text{-}Catechol\text{-}Fe\equiv] + 2H_2O \qquad (4.15)$$

The release of water causes little change to the acidity of the adsorbed water layer.

The assignment of the spectral features in Figure 4.14 was done by a comparison with those collected for aqueous catechol as a function of pH, with and without soluble iron [93]. Under dry conditions, the data suggests the formation of monodentate catechol-Fe complexes that are hydrogen bonded to the neighboring sites. At 30% RH, surface catechol gives rise to spectral features closely resembling those reported for catechol on the surface of hematite [94], and goethite [95] particles at neutral-to-basic pH. In general, a comparison of spectra of surface species with the aqueous phase spectra collected at known pH is useful in inferring the interfacial pH range.

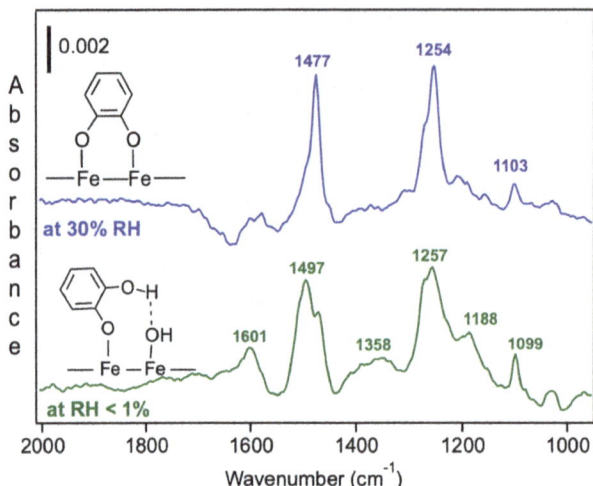

Figure 4.14: DRIFTS absorbance spectra of catechol vapor uptake on solid hematite nanoparticles (1% wt./wt. in diamond powder) under dry (bottom, RH < 1%) and humid (top, 30% RH). Reference spectrum used is that collected for hematite nanoparticles prior to introducing catechol vapor. Reproduced with permission from reference [92] under the Creative Commons Attribution 3.0 Unported License, © The Royal Society of Chemistry, 2015.

4.3 Reactions at the gas/liquid interface

Studying reactions at the gas/liquid interface are relevant to the surfaces of oceans, lakes, cloud and fog droplets, and atmospheric aerosol particles. Samples used in laboratory studies for probing chemistry at the gas/liquid interface are either thin water/solution films on substrates, microdroplets, or simply a beaker/petri dish with liquid. The experimental setup is designed to probe the gas/liquid interface with minimum interference from the bulk gas or liquid. As detailed in Chapter 3 on the hygroscopic properties of atmospheric aerosol particles, there are molecular-level differences in water structure at the gas/water interface versus the bulk. These differences affect the rates and mechanisms of interfacial (photo)chemical reactions [96, 97]. The following paragraphs describe the results from selected studies to illustrate this difference in chemical reactivity.

Tahara and co-workers [98] used the surface-specific ultra-fast vibrational sum frequency generation (VSFG) to study the photochemical dissociation of phenol to a hydrated electron and a hydronium ion through an electronically excited state at the air/water interface. Their results showed that the photoionization reaction of phenol proceeded 10^4 times faster at the water surface than in the bulk aqueous phase (upon irradiation with photons with the same energy). The species at the interface experience diminished solvation, relative to being in bulk aqueous solution, and that reduction in interactions with the solvent lowered the dissociative electronic state, relative to the optically accessible electronic state, creating a smaller barrier for dissociation and a correspondingly faster reaction [96, 98]. Through indirect evidence, Nissenson et al. [99, 100] examined the water-cage effect on the photolysis of NO_3^- and $FeOH^{2+}$ at the air/water interface and found that decreased solvation led to increases in the overall quantum yields for dissociation at the interface, compared to the bulk phase.

As reviewed by Zhong et al. [97], the molecular dynamic simulations of reactions at the aerosol water surface confine the atmospheric species into a specific orientation. The hydrophilicity of these atmospheric species or the hydrogen bonding interactions between them and interfacial water explain this observation, which increases the rate of reactions with the gas phase oxidants (for example), relative to that in the gas phase. In some cases, a different pathway is provided for other types of reaction products not observed in the gas phase. For example, Rossignol et al. [101, 102] showed that the direct photolysis of nonanoic acid – a fatty acid – occured at the air/water interface, leading to the formation of oxidized products in the gas phase and macromolecular products in water. This processing pathway resulted in products that are different than fatty acid oxidation by OH radicals.

Molecular dynamics (MD) simulations were combined with quantum mechanical /classical approaches to describe the ground and excited electronic states of formaldehyde (HCHO) at the air/water interface [103]. Despite its relatively high solubility in bulk water, Figure 4.15 shows that HCHO exhibits a preference for the air/water interface with respect to the bulk by roughly 1.5 kcal mol^{-1}, which was attributed to the decrease of favorable solvent-solvent interactions and solvent entropy accompanying the bulk solvation process. Additional examples are provided in the review by Zhong et al. [104].

Using a vertical wetted wall flow tube (VWWFT) reactor, Gilgorovski and co-workers [105] assessed changes in the heterogeneous ozone uptake to thin water films containing acetosyringone (ACS), a representative methoxyphenol compound, that mimic cloud droplets and deliquesced aerosol particles with high ionic strength. They reported that the reaction kinetics strongly depended on the acidity of the phenolic group of ACS that was affected by increasing the salt (Na_2SO_4) concentration because it decreased the dielectric constant of water and reduced its polarity. At pH 3, a sharp increase was observed in the uptake coefficients of O_3 on aqueous ACS, from $\gamma = (1.4 \pm 0.4) \times 10^{-7}$, in the absence of salt, to $\gamma = (1.2 \pm 0.01) \times 10^{-6}$ at $[Na_2SO_4] = 0.3$ mol L^{-1}.

In a set of experiments probing the gas/aqueous interface, Fenton oxidation of gaseous isoprene was reported by Colussi and co-workers using acidic $FeCl_2$ aqueous microjets [106]. In these experiments, the reactant gaseous streams – isoprene and H_2O_2 – intersected the droplets containing iron for about 10 μs, and the products were analyzed *in situ* via online electrospray ionization-mass spectrometry (ESI-MS). The results indicated the formation of dissolved ˙OH, H_2O_2, HO$_2$˙, and O_2 that led to oxidation products in the condensed phase – mostly polyols and carbonyls.

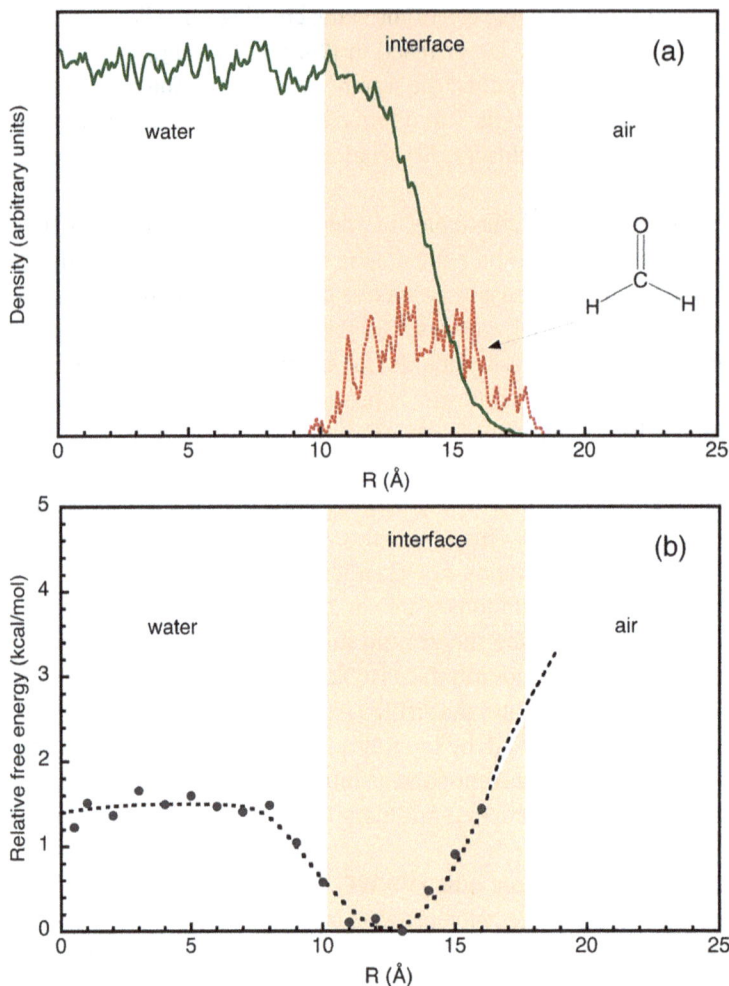

Figure 4.15: Preference of the HCHO molecule for the air/water interface: (a) molecular densities of water (green line) and formaldehyde (red line) from a simulation with formaldehyde freely moving at the interface, (b) bulk/interface free energy profile calculated with the umbrella sampling method. R = 0 corresponds to the center of the simulation box. Figure and caption were reproduced with permission from reference [103], © The American Chemical Society, 2012.

The oxidation process was initiated by OH additions to protonated isoprene oligomer homologues. This iron-driven pathway for VOC oxidation and formation of SOA was reported to be potentially important during day- and night-time.

For studying reactions on single microdroplets, techniques based on optical-tweezers were developed [107, 108]. Studies demonstrating the capabilities of coupling the optically-tweezed aerosol droplets with other techniques, such as Raman spectroscopy, allowed for direct measurements of pH and differentiating the different aerosol morphologies [108]. For example, Chang et al. [109] studied the heterogeneous oxidation of single aqueous ascorbic acid aerosol particles with gas phase O_3 as a function of pH and ionic strength utilizing aerosol optical tweezers with Raman spectroscopy. They reported that the reaction followed the Langmuir-Hinshelwood mechanism, and that the observed aerosol reaction kinetics were dominated by liquid phase diffusion, with the bimolecular rate constants strongly dependent on pH.

4.4 Reactions at the liquid/solid interface

The reactions at the liquid/solid interface in atmospheric aerosol particles are relevant to insoluble inclusions that include dust particles composed of metal oxides, and clay minerals or soot (see Figures 1.5 and 2.2B). As highlighted in Chapter 3, the structure of water at the liquid/solid interface is different from that of bulk water. At the liquid/solid interface, structured hydrogen bonding and layering of ions near the interface take place, particularly near charged surfaces. The techniques used in probing the nature and reactivity of liquid/solid interfaces at the molecular level were reviewed earlier [110]. This section covers the fundamentals of chemical processes at the liquid/solid interface and selected examples of heterogeneous processes, relevant to atmospheric aerosol particles, which include dissolution of metal oxides, complexation of water-soluble organic and inorganic compounds, and catalyzed surface oligomerization/polymerization reactions. Heterogeneous photochemistry at the liquid/solid interface is described in Section 4.5.

4.4.1 Fundamentals of chemical processes at the liquid/solid interface

The water/metal oxide interface has been studied extensively in surface science under vacuum [111] in the aqueous geochemistry literature because the reactivity of the interface is affected by the nature of the first few layers of water in close proximity to the metal oxide (see references [5, 6, 112–117] and details in Chapter 3). Briefly, under aqueous conditions, metal oxide surfaces are hydroxylated [3] and interact not only with water molecules but also with soluble ions, such as hydronium ions, metal cations, hydroxides, nitrates, sulfates, carbonates, and soluble organics. The chemistry that occurs at the surfaces of amphoteric oxides is dominated by the acid–base and the ligand exchange interactions. The acid–base behavior of surface hydroxyl groups is illustrated in the following equation:

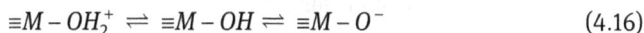

$$\equiv M - OH_2^+ \rightleftharpoons \ \equiv M - OH \rightleftharpoons \ \equiv M - O^- \tag{4.16}$$

where $\equiv M$ denotes the metal cation (e.g., Al(III), Fe(III), Ti(IV)), and $\equiv M - OH_2^+$, $M - OH$, and $\equiv M - O^-$ are protonated, neutral, and deprotonated sites, respectively, which form depending on the pH of bulk water and give rise to surface charge. The

pH_{pzc} is the pH at which the surface is neutral because of the dominance of *M-OH* sites or when the surface coverage of $\equiv M - OH_2^+$ sites equals $\equiv M - O^-$ sites [118, 119]. For the oxides in atmospheric dust, MgO, α-Al$_2$O$_3$, α-Fe$_2$O$_3$, TiO$_2$, and SiO$_2$, the pH_{pzc} are around 12, 9, 9.1, 6, and 2, respectively, depending on the method of preparation [119]. For example, the acid-base behavior at the water/SiO$_2$ interface was studied using X-ray photoelectron spectroscopy [120] and second harmonic generation [121, 122], where the pK_a values of surface groups and the effect of electrolytes on surface charge were quantified. The dissolution rates of metal oxides increase below pH_{pzc} with decreasing pH, and increase with increasing pH in alkaline media [6]. The pH_{pzc} is equivalent to the solution isoelectric point (pH_{isep}) at which the concentration of the positively charged species equals that of the negatively charged ones. However, the actual values of pH_{isep} and pH_{pzc} for the same metal cation are different due to differences in the solvation of surface species versus those in solution and the differences in the coordination shell around the metal in each case.

The surface interactions of hydrated anions, cations, oxyanions, and organic functional groups with the above sites occur via ligand exchange reactions [123, 124]. The kinetics of these reactions and the thermodynamic stability of the surface complexes depend on pH, ionic strength, and temperature. The organic functional groups interact with metal oxides and minerals via a number of mechanisms, such as chelation, ligand exchange, electrostatic, and cation bridging [125]. These surface processes change the surface charge, hydrophobicity, interfacial water structure, and dissolution of the oxide/mineral phase. Quantifying the kinetics and thermodynamics of these surfaces processes under various environmentally relevant conditions is of great importance to developing accurate surface complexation models that take into account the surface charge [126, 127]. These models can predict surface properties under different scenarios, which provide insights into the dissolution behavior and reactivity as detailed in the following sections.

4.4.2 Metal oxide dissolution relevant to atmospheric dust

Understanding the mechanism of atmospheric dust dissolution [128–130] under aerosol and cloud conditions is of importance because this process increases the concentration of the dissolved transition metals, such as iron, which can catalyze several (photo)chemical reactions. Also, the deposition of aged dust on oceans contributes to the pool of dissolved iron, which is considered an essential micronutrient for phytoplankton at the sea surface microlayer (SML) [128–132].

The dissolution mechanisms and kinetics of iron (oxyhydr)oxides, such as hematite, goethite, and ferrihydrite [88–90] were extensively investigated as summarized in reference [92], because of their natural and industrial importance. Iron (oxyhyr) oxide nanoparticles, such as hematite and goethite, have exposed reactive surface planes and a large density of defect sites [133–135]. The main two mechanisms that lead to iron release from dust and iron (oxyhydr)oxides are proton- and ligand-promoted dissolution (see references [134, 136, 137] and the references therein). Laboratory studies from nearly four decades of research into these mechanisms showed that a number of variables play a role, namely, pH, particle size, degree of crystallinity, presence of solar radiation, and adsorption mode of Fe-organic complexes. In general, the highest rates of dissolution occur under acidic conditions (pH < 4) in the presence of solar radiation and oxalate, with nanometer size and amorphous iron-containing particles. These reaction conditions are commonly found in atmospheric aerosol particles as well as in fog and cloud droplets, where soluble and insoluble iron species catalyze several chemical processes [92, 138, 139]. Link et al. [140] quantified the amount of dissolved iron from Arizona Test Dust (AZTD) and from hematite nanoparticles (HEM) at pH 1, relevant to aerosol conditions in the dark, with and without organic and inorganic ligands. Micron-size AZTD particles contain mainly crystalline aluminosilicate clay minerals, and hence are structurally different from iron (oxyhyr)oxides [137]. For example, AZTD particles that are commercially available from Powder Technology Inc. are nominally 0–3 µm in size and contain $3.6 \pm 0.2\%$ (wt./wt.) Fe. Its mineralogy is found to be muscovite (33.4%), quartz (30.7%), albite (10.9%), kaolinite (9.1%), sanidine (7.8%), and calcite (5.4%) using powder X-ray diffraction. Table 4.3 lists the values of dissolved total iron (DFe) for all the samples studied. The values show that for AZTD, the addition of the ligands, pyrocatechol (PC), oxalic acid (Ox), and ammonium sulfate (AS), did not result in significant leaching of iron to the solution, compared to the control sample (no ligands). In the case of hematite nanoparticles, the presence of Ox nearly doubled the amount of iron released into the solution. Since acid-promoted dissolution of iron-containing materials is a surface process, it is imperative to normalize DFe values to the surface area of the particles in the slurries. The third column in Table 4.1 lists the ratio of the DFe mass, relative to the surface area of the solid sample (expressed as DFe_{SA}). This normalization process shows statistically significant higher values for HEM than AZTD, for the samples containing oxalate. For example, DFe_{SA} is 0.3 ± 0.03 for HEM-Ox versus 0.09 ± 0.01 for AZTD-Ox. The values of DFe_{SA}, using PC as a ligand, were comparable between HEM and AZTD, and slightly lower for AZTD using AS. These results suggest that the nanometer size range of the hematite particles affected the oxalate-promoted dissolution mechanism to a higher extent than the micron-sized AZTD particles.

Table 4.3: Quantification of the dissolved iron using ICP-MS, following the filtration of AZTD and hematite samples reaction with pyrocatechol (PC), oxalic acid (Ox), and ammonium sulfate (AS) at pH 1. Table was reproduced with permission from reference [140], © The American Chemical Society, 2020.

Sample	[Fe(aq)] (mg L^{-1})	DFe$_{SA}$ (g m^{-2}) × 100*
AZTD-control	30 ± 3	0.12 ± 0.01
AZTD-PC	20 ± 2	0.08 ± 0.01
AZTD-Ox	22 ± 2	0.09 ± 0.01
AZTD-AS	19 ± 2	0.07 ± 0.01
AZTD-Ox-PC	23 ± 2	0.09 ± 0.01
AZTD-AS-PC	22 ± 2	0.09 ± 0.01
HEM-control	12 ± 2	0.09 ± 0.01
HEM-PC	14 ± 2	0.11 ± 0.01
HEM-Ox	41 ± 4	0.31 ± 0.03
HEM-AS	18 ± 2	0.14 ± 0.01
HEM-Ox-PC	39 ± 4	0.30 ± 0.03
HEM-AS-PC	19 ± 2	0.14 ± 0.01

Notes: * Weight ratio of the dissolved Fe (g) per surface area of solid sample (m^2).

4.4.3 Complexation of soluble inorganic and organic ligands

As stated above and in the following sections, complex formation between metal oxides and soluble inorganic and organic ligands changes the surface properties, dissolution rate, and overall (photo)chemical reactivity. Several books and review articles focused on the type of complexes that form between these ligands and the metal oxides common in atmospheric dust from spectroscopic experiments and theoretical calculations [91, 115, 141–145]. In general, hydrated halides and alkali, and alkaline metal cations, such as Cl$^-$, Br$^-$, I$^-$, K$^+$, Na$^+$, and Mg^{2+} form outer-sphere complexes with most metal oxides of atmospheric relevance.

Because surfaces of metal oxides have charges, the layering of interfacial electrolyte ions, observed in modeling studies [146] and experimental studies [147–149], plays a role in affecting the electrostatic adsorption, particularly for systems containing weakly-bonded outer-sphere complexes. For examples, Situm et al. [150, 151] reported the formation of outer-sphere oxalate complexes on hematite nanoparticles. Varying the electrolyte concentrations, such as KCl, affected the amount of surface oxalate and its adsorption kinetics. As discussed in Situm et al. [150, 151], increasing amounts of K^+(aq) at the interface resulted in a positive charge buildup at the interface, which stabilized the oxalate surface complexes. Cryogenic X-ray photoelectron spectroscopy (XPS) was utilized to gain a molecular-level picture of electrolytes interactions with hematite colloids in aqueous solutions [147–149], which was also predicted using MD simulations. XPS-derived data from *frozen* hematite pastes in 50 mM electrolyte solutions (pH 2 – 11) was used to quantify the surface loadings of anions and cations: $F^- > I^- \approx Cl^- > Br^-$ and Na(F) > Na(I) > Na(Br) > Na(Cl) [149]. The MD calculations at 300 K with NaCl, CsCl, and CsF revealed the presence of a structured interfacial region, resulting from the strong interaction of water with the goethite surface [146]. A buildup of positive charges near the surface was found because of cations adsorption, leading to an accumulation of anions in the next few angstroms. For comparison with deliquesced salt surfaces, XPS experiments and MD predictions at room temperature show enhancement of Br^- and I^- near the interface, compared with Cl^- and F^- [2].

Complex formation of oxyanions, such as nitrate (NO_3^-), sulfate (SO_4^{2-}), and carbonate (CO_3^{2-}) on metal oxides of atmospheric relevance have also been investigated, particularly Al- and Fe-(oxyhyr)oxides [142, 152–155]. Table 4.4 lists the structures of these complexes on metal oxides as identified in experimental infrared spectroscopy studies and theoretical calculations. These anions form inner-sphere complexes over a wide pH range because of the high pH_{pzc} of Al- and Fe-(oxyhyr)oxides. These complexes could be protonated mono- or bi-dentate, depending on the pH of the solution and surface loading. These types of complexes show less sensitivity to electrolyte concentrations under aerosol and cloud pH. For example, SO_4^{2-} forms monodentate inner-sphere complexes with hematite under acidic conditions [154], which is considered weaker than the bidentate and bridging adsorption modes [134, 136], and hence does not enhance dissolution of iron oxides, compared to proton-promoted dissolution. As reviewed by Pincus et al. [156], the adsorption of oxyanions often takes place in the presence of other ligands that have an affinity to metal oxide surfaces. Hence, studying competitive adsorption processes in multicomponent systems is environmentally more relevant than single component systems.

Organic ligands with functional groups of high affinity to complexing metals have also been studied to identify their surface complexes. For example, Plata et al. [162] investigated the reactivity of acetaldehyde with TiO_2 surfaces and identified the role of surface planes and type of surface complexes. The adsorption of tannins on aluminum oxide was found to affect the electron exchange capacities of dissolved and sorbed humic acid [163]. The carboylate groups in natural organic matter were found to exhibit dependency on the exposed surface plane of TiO_2 and Fe_2O_3 nanoparticles [164]. Because of their strong chelating affinity to iron and their role in promoting iron dissolution, the adsorption of soluble organic ligands has also received much attention to identify the type of surface complexes that form and quantify the strength of binding [134, 136]. These organic ligands include oxalate and other dicarboxylic acids and phenolic compounds. Oxalate was found to

Table 4.4: Structure of oxyanion surface complexes on metal (oxyhydr)oxides. Abbreviations: OS = outer-sphere, MM = monodentate mononuclear, BB = bidentate binuclear, BM = bidentate mononuclear, MB = monodentate binuclear.

Anion	Structure	Reference
NO$_3^-$	(OS)　(MM)　(BB)　(BM)　(MB)	[157]
SO$_4^{2-}$	(OS)　(MM)　(BB)	[154, 158]
CO$_3^{2-}$	(MM)　(BM)　(protonated BM)　(BB)	[159–161]

form bidentate mononuclear complexes with surface Fe under acidic conditions, which labilizes surface Fe-O bonds and acts as an electron bridge, leading to the enhanced release of Fe to the solution [136]. In the case of pyrocatechol, the formation of bidentate binuclear complexes [94, 151] with iron oxide surfaces under acidic conditions does not enhance the dissolution of iron compared to proton-promoted dissolution. This observation is explained by the unfavorable activation energy associated with the detachment of two metal cations [165, 166]. Pyrocatechol was found to enhance iron dissolution from hematite particles under neutral to basic conditions [94].

4.4.4 Catalyzed surface oligomerization/polymerization

In the soil chemistry literature, catechol was found to undergo catalytic abiotic oxidation on the surfaces of iron and manganese oxides under *basic* pH (pH > 8). Figure 4.16 shows a scheme for the three-step mechanism in the formation of poly-catechol as per Colarieti et al. [167]. Steps 1 and 2 are heterogeneous reactions, which involve partial oxidation of catechol due to an electron transfer to the surface iron and release of Fe(II) to the solution.

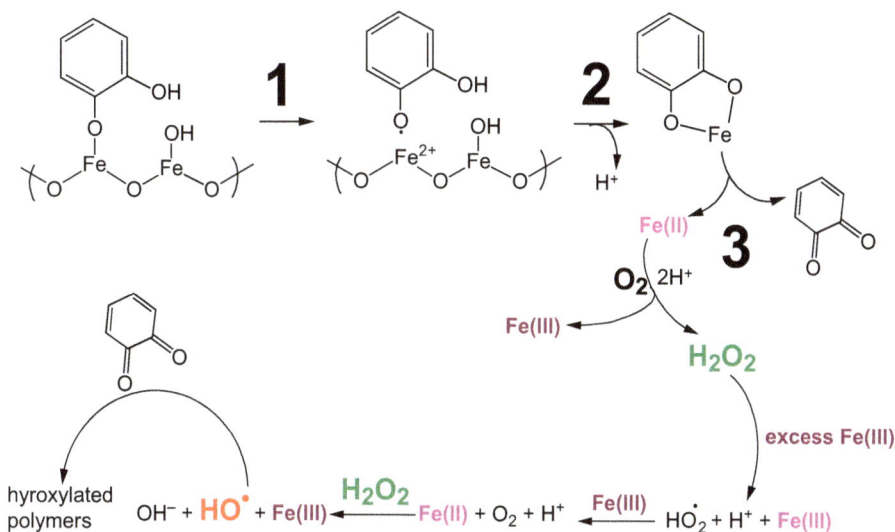

Figure 4.16: A three-step mechanism for abiotic oxidative polymerization of catechol on the surface of iron oxide per Colarieti et al. [167], and Larson and Hufnal Jr. [168]. Different colors are used to highlight the oxidation state of the iron in the respective step. A larger size for OH and H_2O_2 is used to highlight the reactive oxygen species generated in the reaction. Reproduced from ref. [140] with permission from the American Chemical Society, © 2020.

The reduced Fe(II) species in the solution further complexes to catechol, which undergoes homogeneous oxidation as per step 3, in the presence of dissolved oxygen. This reaction step involves cycling between Fe(II) and Fe(III) species and the production of the reactive oxygen species, H_2O_2, and OH radicals, as intermediates [168].

Under *acidic* conditions relevant to atmospheric aerosol pH, the same chemistry takes place, albeit over days of reaction time. The following two examples highlight this point in more detail, and show simulated acid-driven dissolution of freshly-emitted dust and its reactivity with catechol using hematite [169] and AZTD [140]. Figure 4.17 shows digital images of the hematite slurry following 10 days of mixing at pH 1. Also shown are images of the filtrates, before and after the addition of catechol and guaiacol, in addition to the filters after 1 h of reaction.

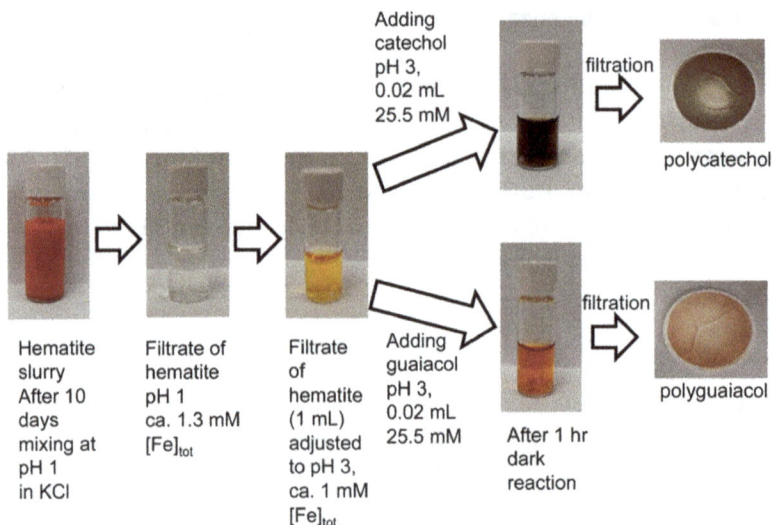

Figure 4.17: Formation of polycatechol and polyguaiacol from a reaction with dissolved iron from acid-promoted hematite dissolution. Reproduced with permission from reference [169], © American Chemical Society, 2015.

Using AZTD, Link et al. [140] studied its reaction with dissolved catechol as a function of time and pH in the dark. The left and right panels of Figure 4.18 show photographs of AZTD control (C, no catechol added) and reaction (R, with catechol) slurries over a period of 14 days, and that of dry particles after filtration. Photographs of the control AZTD slurries (Figures 4.18) showed little change in color over the 14 d reaction time. However, the color of the particles on the filters from the *reaction* vials appeared darker compared to that from the *control* vials with no catechol. Similar studies were conducted using hematite nanoparticles under acidic pH, where progressive darkening in the color of the slurry containing catechol with time was observed (see Supporting Information in ref. [140]). The darker color shown of the dried filters obtained after 14 days of simulated acid processing suggested that the adsorption of catechol to iron-containing materials slowly changes the surface chemical composition of the particles. Figure 4.19 shows the scanning transmission electron microscopy (STEM) images of AZTD particles with catechol (top) and unreacted (bottom), following nearly two weeks of reaction at pH 1. The STEM images were coupled with energy dispersive X-ray spectroscopy (EDS) elemental mapping to show the distribution of carbon, oxygen and iron in the samples. The iron signals are concentrated in some areas – brighter areas correspond to higher relative amounts than dimmer areas. The carbon signals in Figure 4.19a for particles reacted with catechol are not uniformly distributed across the image, with some areas having larger amount of detected carbon than the others. The C atomic % is higher for AZTD-Cat at 4.5%, compared to undetectable for the control AZTD (no catechol in Figure 4.19b). Other STEM

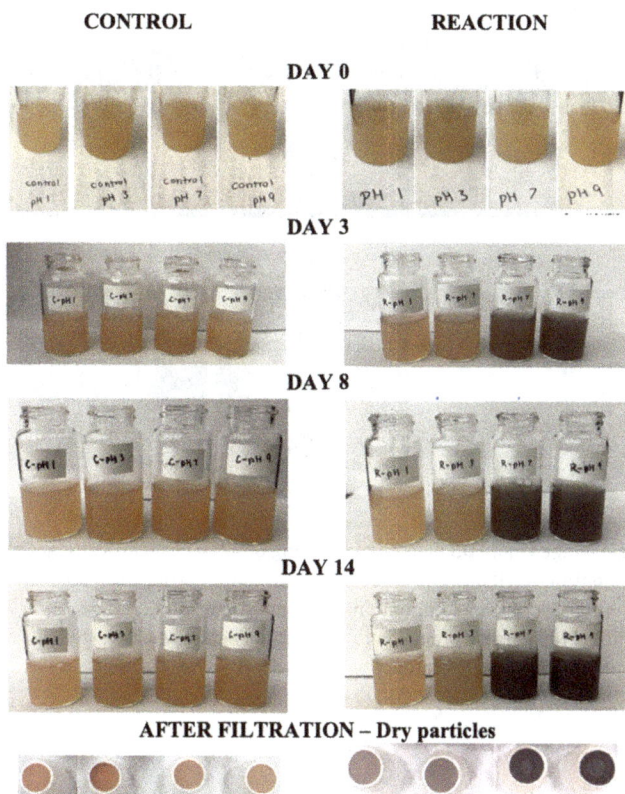

Figure 4.18: Digital photographs of control (C, left) and reaction (R, right) vials containing AZTD as a function of pH and simulated atmospheric aging time. The pH values shown in the label of each vial refers to the starting pH of the slurry. At the end of the day 14, the slurries were filtered and the corresponding filters are shown below each vial. The concentration of DFe in the filtrates of C-pH 1, R-pH 1, and R-pH 3 were 9.5, 12.2, and 2.3 mg L^{-1}, respectively. Values of DFe for the other filtrate solutions were below the detection limit of the instrument (0.25 mg L^{-1}). Reproduced from ref. [140] with permission from the American Chemical Society, © 2020.

images in reference [140] showed that there is some co-localization of the iron and the carbon signals, suggesting that the iron content in the dust particles might be anchoring the polymeric network of polycatechol, which upon formation, might encapsulate the dust particle.

In summary, the change in the chemical composition of iron-containing particles, driven by oxidative polymerization of catechol, changes the optical properties, mixing state, and morphology of the particles.

(a) AZTD-Cat

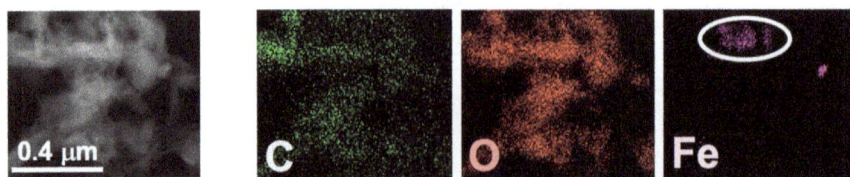

(b) AZTD only

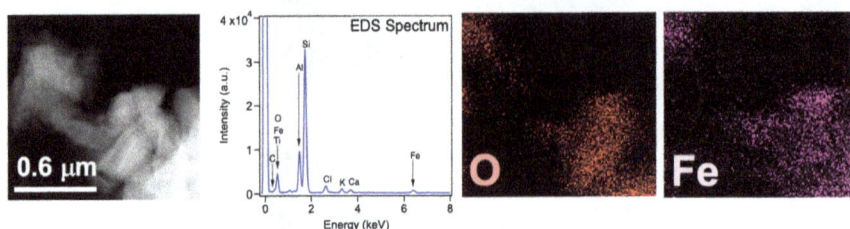

Figure 4.19: Representative STEM images and EDS elemental mapping of AZTD particles with (a) polycatechol (AZTD-Cat) and (b) control (AZTD only, no catechol in solution). Slurries used for the images are shown in Figure S3 in the Supporting Information. The images labeled C, O, and Fe represent elemental maps for the corresponding atoms. In (b), no carbon signal was detected in the EDS spectrum, hence, no carbon map was generated. Reproduced from ref. [140] with permission from the American Chemical Society, © 2020.

4.5 Heterogeneous photochemistry

Current cloud chemistry models contain thermodynamic and kinetic parameters from known photochemical reactions in the bulk aqueous phase [170]. These reactions do not necessarily and accurately represent the photochemical reactions occurring at the surface of aerosols or at atmospherically-relevant surfaces [2]. Heterogeneous photocatalysis on inorganic and organic materials of atmospheric relevance are known to generate ROS (See references [26, 28, 171, 172] for comprehensive reviews). Questions exist on the relative efficiency of *interfacial photochemistry* in the presence of few layers of adsorbed water compared to that in bulk liquid water under photon fluxes that simulate the solar flux. The photocatalytic reactivity of metal oxides was demonstrated in several studies that involve the gas phase and the dissolved species of importance to atmospheric chemistry (see reviews [26, 54, 139, 173–175]). For example, photoenhanced uptake of O_3 [176] and NO_2 [177] was reported on mineral dust. The mechanisms include the photogeneration of electron-hole pairs, which serve as the driver for charge transfer reactions with adsorbed water, O_3, and NO_2. In the case of VOC, photoenhanced uptake of formaldehyde [178, 179] and short-chain alcohols [180] was reported on TiO_2, Fe_2O_3, dust, and volcanic ash. The ˙OH radical that forms from the reaction of adsorbed water with photogenerated electron-hole pairs seems to explain these experimental observations. The photo uptake was reported to be RH-dependent, as in the case of formaldehyde [179], where the uptake coefficient reached a maximum at 30% RH, and then declined at higher values due to site blockage by adsorbed water. Also, nitrite and chloride anions get oxidized from reactions with holes, which in the presence of aromatic compounds can lead to nitration and chlorination of the benzene ring [174]. Some of the organic content in aerosol particles is photoactive, particularly carbonyl, and aromatic containing compounds [26, 60, 171, 181]. The following sections highlight several relevant studies in these areas.

4.5.1 TiO$_2$

The surface properties and photocatalytic reactivity of TiO$_2$ were the subject of two comprehensive reviews [172, 182]. Photocatalytic oxidation of adsorbed organics, such as HCHO by TiO$_2$, was reported to take place, forming CO, CO$_2$, and H$_2$O, with HCOOH/HCOO$^-$ as the intermediates and/or products. Because of this chemistry, regeneration of surface sites takes place, leading to enhanced uptake of VOC under illumination, compared to dark conditions [82]. In a study by Chu et al. [183], the heterogeneous reactions of SO$_2$ in the presence of NO$_2$ and C$_3$H$_6$ on TiO$_2$ were investigated under dark and UV-vis irradiation conditions. Figure 4.20 shows the amount of surface sulfate that forms under various experimental conditions. Sulfite formation (instead of sulfate) was observed when SO$_2$ reacted with TiO$_2$ in the dark per the following equations:

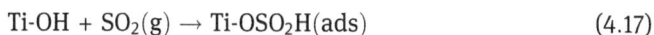

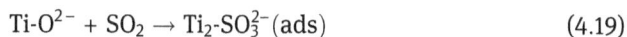

$$\text{Ti-OH} + \text{SO}_2(\text{g}) \rightarrow \text{Ti-OSO}_2\text{H(ads)} \tag{4.17}$$

$$2\text{Ti-OH} + \text{SO}_2(\text{g}) \rightarrow \text{Ti}_2\text{-SO}_3 \cdot \text{H}_2\text{O(ads)} \tag{4.18}$$

$$\text{Ti-O}^{2-} + \text{SO}_2 \rightarrow \text{Ti}_2\text{-SO}_3^{2-}\text{(ads)} \tag{4.19}$$

Surface sulfate was detected upon flowing ppb levels of NO$_2$ because of two potential pathways that can oxidize S(IV) to S(VI): one pathway leads to the formation of HNO$_3$ and HONO upon NO$_2$ adsorption on TiO$_2$ through N$_2$O$_4$ as a surface intermediate, and another pathway that requires water is similar to the aqueous phase oxidation of SO$_2$ by NO$_2$. The presence of C$_3$H$_6$ in the dark did not result in sulfate formation on the surface of TiO$_2$, whereas it suppressed the sulfate formation when NO$_2$ was present because of the potential oxidation of surface C$_3$H$_6$ to HCHO and H$_3$CCHO by surface RNS, namely, NO$^+$NO$_3^-$. Under UV-vis irradiation, the coexistence of NO$_2$ and/or C$_3$H$_6$ significantly suppressed the sulfate formation due to competition for surface ROS and other active sites that form on TiO$_2$, according to the following reactions:

$$TiO_2 + h\nu(\lambda < 387\text{nm}) \rightarrow e^- h^+ \rightarrow e^- + h^+ \qquad (4.20)$$

$$O_2 + e^- \rightarrow O_2^- \qquad (4.21)$$

$$H_2O + h^+ \rightarrow \cdot OH + H^+ \qquad (4.22)$$

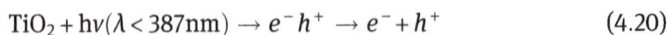

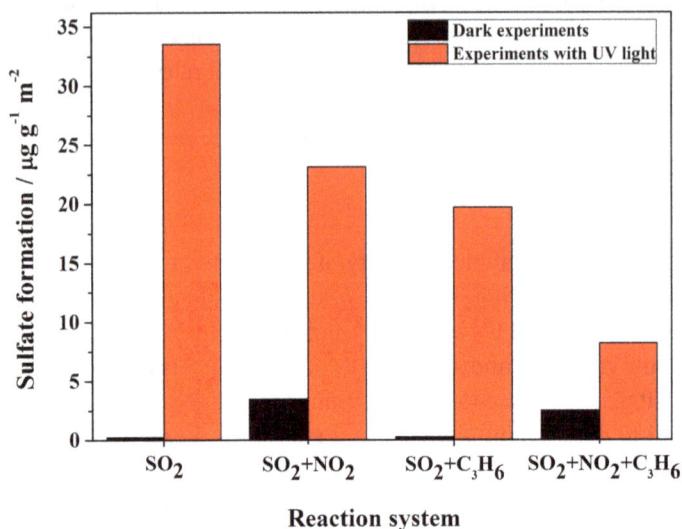

Figure 4.20: Ion chromatography results of the amounts of sulfate (product per unit mass divided by surface area of sample) formed on the surface of TiO$_2$, after reaction with SO$_2$, SO$_2$ + NO$_2$, SO$_2$ + C$_3$H$_6$, and SO$_2$ + NO$_2$ + C$_3$H$_6$ in experiments under dark conditions or with UV–Vis light. Since formaldehyde was added to inhibit the oxidation of sulfite to sulfate in the solution, there is a possibility that hydroxymethanesulfonate (HMS) would be generated in the solution and would be measured as sulfate. Figure and caption were modified from reference [183] under the Creative Commons Attribution 4.0 License (CC-BY), © The Author(s), 2019.

4.5.2 Fe$_2$O$_3$/FeOOH

The heterogeneous photochemistry of iron (oxyhydr)oxides has been reviewed earlier [92]. Briefly, these materials have band gaps in the visible region of the electromagnetic spectrum: hematite is around 2.2 eV (565 nm) [184] and goethite ranges from 2.1–2.5 eV (592–497 nm) [89]. These values were found to increase with decreasing particle size [185, 186]. Hence, formation of electron-hole pairs followed by Fe(II) species results from the band gap excitations with UV photons with energies > 2.1 eV. In the presence of inorganic gases, such as nitrogen and sulfur oxides, the photoreactivity of iron (oxyhydr)oxides was shown to be important for renoxification and oxidation of sulfur dioxide [187]. The formation of electron-hole pairs in the presence of organics can initiate redox chemistry through ligand to metal charge transfer (LMCT) to surface $\equiv$Fe(III), leading to the formation of the reduced iron sites, $\equiv$Fe(II) [174].

These species then desorb to the bulk phase, resulting in iron dissolution in a process called photoreductive dissolution. In the case of hematite, the photochemical mechanism was explained with the following equations, which can be generalized for other iron (oxyhydr)oxides [188]:

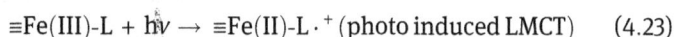

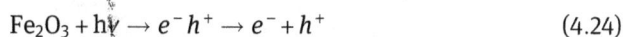

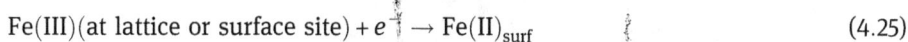

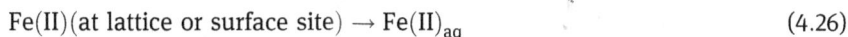

$$\equiv Fe(III)\text{-}L + h\nu \rightarrow \ \equiv Fe(II)\text{-}L \cdot {}^+ \ (\text{photo induced LMCT}) \quad (4.23)$$

$$Fe_2O_3 + h\nu \rightarrow e^- h^+ \rightarrow e^- + h^+ \quad (4.24)$$

$$Fe(III)(\text{at lattice or surface site}) + e^- \rightarrow Fe(II)_{surf} \quad (4.25)$$

$$Fe(II)(\text{at lattice or surface site}) \rightarrow Fe(II)_{aq} \quad (4.26)$$

4.5.3 Al$_2$O$_3$

Alumina (Al$_2$O$_3$) is regarded as an insulator and hence few studies investigated its photoreactivity, in general. A recent discovery by Leow et al. [189] showed that surface complexation of benzyl alcohol with the Bronsted base sites on Al$_2$O$_3$ reduces its oxidation potential and causes an upshift in its HOMO level for electron abstraction by a photoactive dye. Electron transfer to adsorbed O$_2$ from photoexcited dyes facilitates the overall transfer of protons and electrons from benzyl alcohol to O$_2$. Given the abundance of alumina in mineral dust and organic matter in the gas phase and aerosol particles, there is a need to explore the photoreactivity of alumina in the presence of organics under atmospherically relevant conditions.

The effect of Al$_2$O$_3$ seed particles on SOA formation was investigated by a number of groups [190, 191]. These studies involved the photooxidation of reactants and were not based on the photoreactivity of Al$_2$O$_3$. These studies are included in this section to highlight the role of Al$_2$O$_3$ in providing surfaces for heterogeneous nucleation of SOA, and hence exploring the photoreactivity of these coated Al$_2$O$_3$ particles is related to the above recommendation. Chu et al. [190] reported results from smog chamber measurements using either toluene or α-pinene under NO$_x$- and VOC-limited conditions, and explored the effects of SO$_2$ and NH$_3$ on SOA particle mass. They reported a synergetic promoting effect of Al$_2$O$_3$ and NO$_x$ on SOA formation in heterogeneous reactions, which strengthened the positive relationship between NO$_x$ and SOA formation under "NO$_x$-limited" conditions. Under "VOC-limited" conditions, the synergetic effect reversed the negative relationship between NO$_x$ and SOA formation. The formation of sulfate and nitrate was observed in these experiments, with the latter generated in experiments at lower temperatures than 298 K and in the presence of NH$_3$. In related studies [191, 192], the effect of Al$_2$O$_3$ (Figure 4.21) and Al$_2$(SO$_4$)$_3$ seed particles was used for comparison with H$_2$SO$_4$ particles on SOA formation in toluene/NO$_x$ photooxidation, under SO$_2$ and NH$_3$. They found that under NH$_3$-poor conditions, Al$_2$O$_3$ has an inhibiting effect on SOA formation that decreased with increasing concentration of SO$_2$. Weakly acidic Al$_2$(SO$_4$)$_3$ seeds did not have an obvious effect on SOA formation.

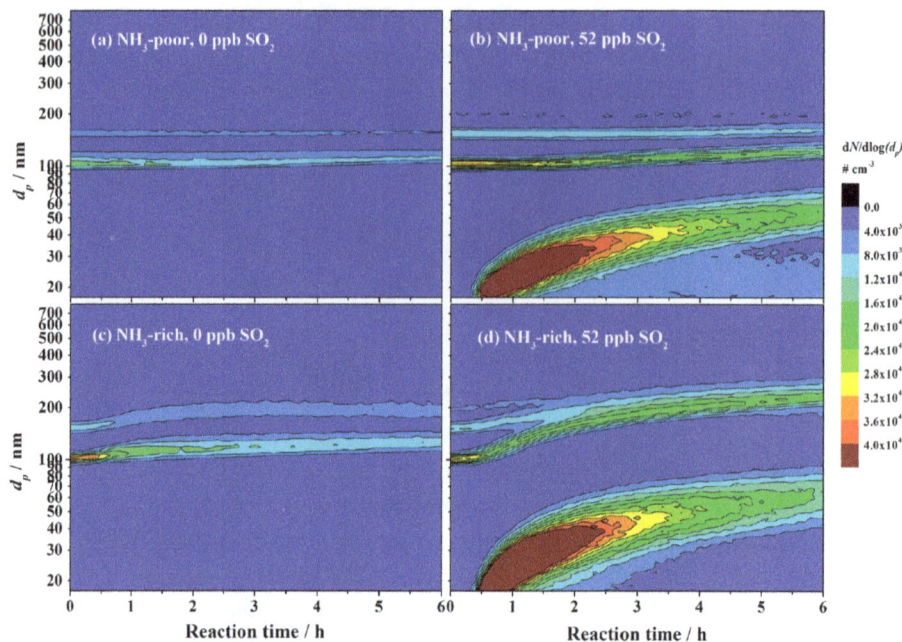

Figure 4.21: Size distributions of the suspended particles, as a function of time, during the reaction in the photooxidation of toluene/NO$_x$ in the presence of Al$_2$O$_3$ seed particles. Figure and caption were reproduced from reference [192] under the CC-BY Attribution 3.0 License, © The Author(s), 2016.

4.5.4 Organics

Atmospheric humic-like substances (HULIS) absorb in the UV-visible region of the actinic flux ($\lambda > 290$ nm) and hence can be involved in direct or sensitized photochemical reactions. While *bulk* photochemical reactions involving humic substances in aquatic systems received much attention [138, 174, 175, 193, 194], fewer studies reported the heterogeneous photochemistry of atmospheric HULIS [26]. This is likely because the efficiency of photoreactivity depends on the amount of aerosol water. Experiments on the hygroscopic properties of HULIS in aerosols showed a continuous water uptake as a function of RH, except for fulvic acid that showed phase separation [195–198]. The structure of water on some of these systems resembled that of water at the interface of polar organic solvents [199]. The amount of water uptake was found to vary depending on the carbon functional groups in fulvic acid samples after phase separation [198]. This means that depending on the size and the water content of the particles, preferential partitioning of HULIS to the surface could become important in impacting their overall surface reactivity. The following key studies illustrate the role of HULIS in heterogeneous photochemistry: (a) fast photosensitized formation of

HONO from photoreaction of NO_2 with humic [60, 200], tannic, and gentisic acid films [201] under dry and humid conditions, (b) the photooxidation and formation of secondary particles from catechol and guaiacol [202], and (c) changes to the functional groups, characteristic of solid tannic acid, and the formation of new carbonyl groups, characteristic of aryl aldehydes and/or quinone, under humid conditions [199]. Moreover, the structure of HULIS contains functional groups capable of complexing iron and these complexes are photoactive as well [92, 171]. Section 4.5.2 listed references for the heterogeneous photochemical reactions with surrogates of HULIS, in the presence of iron, such as catechol and gallic acid.

In summary, as observed in the dark, there are key mechanistic differences in the heterogeneous photochemical reactivity of metal oxides and organics that depend on their hygroscopicity and amount of surface water.

References

[1]   Boucher O. Atmospheric Aerosols: Properties and Impacts. France: Springer; 2015.
[2]   Finlayson-Pitts BJ. Reactions at surfaces in the atmosphere: Integration of experiments and theory as necessary (but not necessarily sufficient) for predicting the physical chemistry of aerosols. Phys Chem Chem Phys. 2009;11(36):7760–7779.
[3]   Tamura H, Mita K, Tanaka A, Ito M. Mechanism of hydroxylation of metal oxide surfaces. J Coll Inter Sci. 2001;243:202–207.
[4]   Boehm HP. Acidic and basic properties of hydroxylated metal oxide surfaces. Discuss Faraday Soc. 1971;52:264–275.
[5]   Stumm W. Chemistry of the Solid-Water Interface. New York: John Wiely & Sons, Inc.; 1992.
[6]   Stumm W. Reactivity at the mineral-water interface: Dissolution and inhibition. Colloids Surf A. 1997;21:143–166.
[7]   Grassian VH. Surface science of complex environmental interfaces: Oxide and carbonate surfaces in dynamic equilibrium with water vapor. Surf Sci. 2008;602:2955–2962.
[8]   Seinfeld JH, Pandis SN. Atmospheric Chemistry and Physics: From Air Pollution to Climate Change. New York: Wiley; 2006.
[9]   Seinfeld JH, et.al. VOCs and Nox: Relationship to Ozone and Associated Pollutants. Rethinking the Ozone Problem in Urban and Regional Air Pollution. Washington, DC: The National Academies Press; 1991. p. 163–186.
[10]   Jacob DJ. Heterogeneous chemistry and tropospheric ozone. Atmos Environ. 2000;34: 2131–2159.
[11]   Finlayson-Pitts BJ, Pitts JN Jr. Chemistry of the Upper and Lower Atmosphere. New York: Academic Press; 2000.
[12]   Fleming ZL, Doherty RM, Von Schneidemesser E, Malley CS, Cooper OR, Pinto JP, et al. Tropospheric ozone assessment report: Present-day ozone distribution and trends relevant to human health. Elem Sci Anth. 2018;6:12–41.
[13]   Usher CR, Michel AE, Grassian VH. Reactions on mineral dust. Chem Rev. 2003;103: 4883–4940.
[14]   Chapleski JRC, Zhang Y, Troya D, Morris JR. Heterogeneous chemistry and reaction dynamics of the atmospheric oxidants, O_3, NO_3, and OH, on organic surfaces. Chem Soc Rev. 2016;45:3731–3746.

[15] Vesna O, Sjogren S, Weingartner E, Samburova V, Kalberer M, Gaggeler HW, et al. Changes of fatty acid aerosol hygroscopicity induced by ozonolysis under humid conditions. Atmos Chem Phys. 2008;8:4683–4690.

[16] Jwl L, Carrascon V, Gallimore PJ, Fuller SJ, Bjorkegren A, Spring DR, et al. The effect of humidity on the ozonolysis of unsaturated compounds in aerosol particles. Phys Chem Chem Phys. 2012;14:8023–8031.

[17] Chu Y, Cheng TF, Gen M, Chan CK, Lee AKY, Chan MN. Effect of ozone concentration and relative humidity on the heterogeneous oxidation of linoleic acid particles by ozone: An insight into the interchangeability of ozone concentration and time. ACS Earth Space Chem. 2019;3:779–788.

[18] Pillar EA, Zhou R, Guzman MI. Heterogeneous oxidation of catechol. J Phys Chem A. 2015;119(41):10349–10359.

[19] Woden B, Skoda MWA, Milsom A, Gubb C, Maestro A, Tellam J, et al. Ozonolysis of fatty acid monolayers at the air–water interface: Organic films may persist at the surface of atmospheric aerosols. Atmos Chem Phys. 2021;21:1325–1340.

[20] Dong X, Fu JS, Huang K, Tong D, Zhuang G. Model development of dust emission and heterogeneous chemistry within the community multiscale air quality modeling system and its application over East Asia. Atmos Chem Phys. 2016;16:8157–8180.

[21] Wang X, Romanias MN, Thevenet F, Rousseau A. Geocatalytic uptake of ozone onto natural mineral dust. Catalysts. 2018;8(263).doi:10.3390/catal8070263.

[22] Nicolas M, Ndour M, Oumar KA, D'Anna B, George C. Photochemistry of atmospheric dust: Ozone decomposition on illuminated titanium dioxide. Environ Sci Technol. 2009;43: 7437–7442.

[23] Woodill LA, O'Neill EM, Hinrichs RZ. Impacts of surface adsorbed catechol on tropospheric aerosol surrogates: Heterogeneous ozonolysis and its effects on water uptake. J Phys Chem A. 2013;117(27):5620–5631.

[24] Tang M, Cziczo DJ, Grassian VH. Interactions of water with mineral dust aerosol: Water adsorption, hygroscopicity, cloud condensation, and ice nucleation. Chem Rev. 2016;116: 4205–4259.

[25] Coates Fuentes ZL, Kucinski TM, Hinrichs RZ. Ozone decomposition on kaolinite as a function of monoterpene exposure and relative humidity. ACS Earth Space Chem. 2018;2:21–30.

[26] George C, Ammann M, D'Anna B, Donaldson DJ, Nizkorodov SA. Heterogeneous photochemistry in the atmosphere. Chem Rev. 2015;115(10):4218–4258.

[27] Ma J, Liu Y, Ma Q, Liu C, He H. Heterogeneous photochemical reaction of ozone with anthracene adsorbed on mineral dust. Atmos Environ. 2013;72:165–170.

[28] Nosaka Y, Nosaka A. Generation and detection of reactive oxygen species in photocatalysis. Chem Rev. 2017;117:11302–11336.

[29] Lasne J, Romanias MN, Thevenet F. Ozone uptake by clay dusts under environmental conditions. ACS Earth Space Chem. 2018;2:904–914.

[30] Simpson WR, Brown SS, Saiz-Lopez A, Thornton JA, Von Glasow R. Tropospheric halogen chemistry: Sources, cycling, and impacts. Chem Rev. 2015;115:4035–4062.

[31] Tobias DJ, Stern AC, Baer MD, Levin Y, Mundy CJ. Simulation and theory of ions at atmospherically relevant aqueous liquid-air interfaces. Annu Rev Phys Chem. 2013;64: 339–359.

[32] Jungwirth P, Tobias DJ. Specific ion effects at the air/water interface. Chem Rev. 2006;106: 1259–1281.

[33] Ghosal S, Hemminger JC, Bluhm H, Mun BS, Hebenstreit ELD, Ketteler G, et al. Electron spectroscopy of aqueous solution interfaces reveals surface enhancement of halides. Science. 2005;307:563–566.

[34] Bartels-Rausch T, Jacobi HW, Kahan TF, Thomas JL, Thomson ES, Abbatt JPD, et al. A review of air–ice chemical and physical interactions (AICI): Liquids, quasi-liquids, and solids in snow. Atmos Chem Phys. 2014;14:1587–1633.

[35] O'Driscoll P, Lang K, Minogue N, Sodeau J. Freezing halide ion solutions and the release of interhalogens to the atmosphere. J Phys Chem A. 2006;110(14):4615–4618.

[36] Clifford D, Donaldson DJ. Direct experimental evidence for a heterogeneous reaction of ozone with bromide at the air-aqueous interface. J Phys Chem A. 2007;111:9809–9814.

[37] Oldridge NW, Abbatt JPD. Formation of gas-phase bromine from interaction of ozone with frozen and liquid NaCl/NaBr solutions: quantitative separation of surficial chemistry from bulk-phase reaction. J Phys Chem A. 2011;115:2590–2598.

[38] Wren SN, Kahan TF, Jumaa KB, Donaldson DJ. Spectroscopic studies of the heterogeneous reaction between O_3(g) and halides at the surface of frozen salt solutions. J Geophys Res. 2010;115:D16309, doi:10.1029/2010JD013929.

[39] Gligorovski S, Strekowski R, Barbati S, Vione D. Environmental implications of hydroxyl radicals (•OH). Chem Rev. 2015;115:13051–13092.

[40] Elshorbany Y, Barnes I, Becker KH, Kleffmann J, Wiesen P. Sources and cycling of tropospheric hydroxyl radicals – an overview. Z Phys Chem. 2010;224:967–987.

[41] Alicke B, Platt U. Impact of nitrous acid photolysis on the total hydroxyl radical budget during the limitation of oxidant production/pianura padana produzione di ozono study in Milan. J Geophys Res. 2002;107(D22):8196, doi:10.1029/2000JD000075.

[42] Stone D, Whalley LK, Heard DE. Tropospheric OH and HO_2 radicals: Field measurements and model comparisons. Chem Soc Rev. 2012;41:6348–6404.

[43] Atkinson R, Arey J. Atmospheric degradation of volatile organic compounds. Chem Rev. 2003;103:4605.

[44] George IJ, Abbatt JPD. Heterogeneous oxidation of atmospheric aerosol particles by gas-phase radicals. Nature Chem. 2010;2:713–722.

[45] Ramasamy S, Nakayama T, Morino Y, Imamura T, Kajii Y, Enami S, et al. Nitrate radical, ozone and hydroxyl radical initiated aging of limonene secondary organic aerosol. Atmos Environ X. 2021;9(100102):1–10.

[46] Heath AA, Valsaraj KT. Effects of temperature, oxygen level, ionic strength, and pH on the reaction of benzene with hydroxyl radicals at the air–water interface in comparison to the bulk aqueous phase. J Phys Chem A. 2015;119(31):8527–8536.

[47] Lam HK, Xu R, Choczynski J, Davies JF, Ham D, Song M, et al. Effects of liquid–liquid phase separation and relative humidity on the heterogeneous OH oxidation of inorganic–organic aerosols: Insights from methylglutaric acid and ammonium sulfate particles. Atmos Chem Phys. 2021;21:2053–2066.

[48] Sjostedt SJ, Abbatt JPD. Release of gas-phase halogens from sodium halide substrates: Heterogeneous oxidation of frozen solutions and desiccated salts by hydroxyl radicals. Environ Res Lett. 2008;3(1-7):045007.

[49] Jacob DJ. Introduction to Atmospheric Chemistry. Princeton, N.J.: Princeton University Press; 1999.

[50] Gentner DR, Jathar SH, Gordon TD, Bahreini R, Day DA, El Haddad I, et al. Review of urban secondary organic aerosol formation from gasoline and diesel motor vehicle emissions. Environ Sci Technol. 2017;51(3):1074–1093.

[51] Khan MAH, Cooke MC, Utembe SR, Archibald AT, Derwent RG, Xiao P, et al. Global modeling of the nitrate radical (NO_3) for present and pre-industrial scenarios. Atmos Res. 2015; 164–165:347–357.

[52] Ng NL, Brown SS, Archibald AT, Atlas E, Cohen RC, Crowley JN, et al. Nitrate radicals and biogenic volatile organic compounds: Oxidation, mechanisms, and organic aerosol. Atmos Chem Phys. 2017;17:2103–2162.

[53] Lu X, Wang Y, Li J, Shen L, Fung JCH. Evidence of heterogeneous HONO formation from aerosols and the regional photochemical impact of this HONO source. Environ Res Lett. 2018;13(114002):1–13.

[54] Ma J, Liu Y, Han C, Ma Q, Liu C, He H. Review of heterogeneous photochemical reactions of NOy on aerosol – A possible daytime source of nitrous acid (HONO) in the atmosphere. J Environ Sci. 2013;25(2):326–334.

[55] Khalizov A, Cruz-Quinones M, Zhang R. Heterogeneous reaction of NO2 on fresh and coated soot surfaces. J Phys Chem A. 2010;114(28):7516–7524.

[56] Al-Abadleh HA, Grassian VH. Heterogeneous reaction of NO2 on hexane soot: A knudsen cell and FT-IR study. J Phys Chem A. 2000;104:11926–11933.

[57] Aubin DG, Abbatt JPD. Interaction of NO_2 with hydrocarbon soot: Focus on HONO yield, surface modification, and mechanism. J Phys Chem A. 2007;111:6263–6273.

[58] Monge ME, D'Anna B, Mazria L, Giroir-Fendlera A, Ammann M, Donaldson DJ, et al. Light changes the atmospheric reactivity of soot. Proc Natl Acad Sci USA. 2010;107:6605–6610.

[59] Spataro F, Ianniello A. Sources of atmospheric nitrous acid: State of the science, current research needs, and future prospects. J Air Waste Manage Assoc. 2014;64(11):1232–1250.

[60] Kleffmann J. Daytime sources of nitrous acid (HONO) in the atmospheric boundary layer. Chem Phys Chem (Mini Review). 2007;8:1137–1144.

[61] Scharko NK, Martin ET, Losovyj Y, Peters DG, Raff JD. Evidence for quinone redox chemistry mediating daytime and nighttime NO2-to-HONO conversion on soil surfaces. Environ Sci Technol. 2017;51(17):9633–9643.

[62] Sander R. Compilation of Henry's law constants (version 4.0) for water as solvent. Atmos Chem Phys. 2015;15(8):4399–4981.

[63] Wang M, Kong W, Marten R, et al. Rapid growth of new atmospheric particles by nitric acid and ammonia condensation. Nature. 2020;581:184–189.

[64] Ault AP, Guasco TL, Baltrusaitis J, Ryder OS, Trueblood JV, Collins DB, et al. Heterogeneous reactivity of nitric acid with nascent sea spray aerosol: Large differences observed between and within individual particles. J Phys Chem Lett. 2014;5(15):2493–2500.

[65] Trueblood JV, Estillore AD, Lee C, Dowling JA, Prather KA, Grassian VH. Heterogeneous chemistry of lipopolysaccharides with gas-phase nitric acid: Reactive sites and reaction pathways. J Phys Chem A. 2016;120:6444–6450.

[66] Chang WL, Bhave PV, Brown SS, Riemer N, Stutz J, Dabdub D. Heterogeneous atmospheric chemistry, ambient measurements, and model calculations of N_2O_5: A review. Aerosol Sci Technol. 2011;45(6):665–695.

[67] Osthoff HD, Sommariva R, Baynard T, Pettersson A, Williams EJ, Lerner BM, et al. Observation of daytime N2O5 in the marine boundary layer during New England air quality study–intercontinental transport and chemical transformation 2004. J Geophys Res-Atmos. 2006;111(D23), doi:10.1029/2006JD007593.

[68] Kim MJ, Farmer DK, Bertram TH. A controlling role for the air–sea interface in the chemical processing of reactive nitrogen in the coastal marine boundary layer. PNAS. 2014;111(11): 3943–3948.

[69] Thornton JA, Kercher JP, Riedel TP, Wagner NL, Cozic J, Holloway WPD, et al. A large atomic chlorine source inferred from mid-continental reactive nitrogen chemistry. Nature. 2010;464: 271–274.

[70] Tham YJ, Wang Z, Li Q, Wang W, Wang X, Lu K, et al. Heterogeneous N_2O_5 uptake coefficient and production yield of $ClNO_2$ in polluted northern China: Roles of aerosol water content and chemical composition. Atmos Chem Phys. 2018;18:13155–13171.

[71] Yu C, Wang Z, Xia M, Fu X, Wang W, Tham YJ, et al. Heterogeneous N_2O_5 reactions on atmospheric aerosols at four Chinese sites: Improving model representation of uptake parameters. Atmos Chem Phys. 2020;20:4367–4378.

[72] Tang MJ, Thieser J, Schuster GL, Crowley JN. Kinetics and mechanism of the heterogeneous reaction of N_2O_5 with mineral dust particles. Phys Chem Chem Phys. 2012;14:8551–8561.

[73] Xia M, Wang W, Wang Z, Gao J, Li H, Liang Y, et al. Heterogeneous uptake of N_2O_5 in sand dust and urban aerosols observed during the dry season in Beijing. Atmosphere. 2019;10 (4):2041-16.

[74] McDuffie EE, Fibiger DL, Dube WP, Lopez-Hilfiker F, Lee BH, Thornton JA, et al. Heterogeneous N_2O_5 uptake during winter: aircraft measurements during the 2015 WINTER campaign and critical evaluation of current parameterizations. J Geophys Res-Atmos. 2018;123(8): 4345–4372.

[75] Knopf DA, Forrester SM, Slade JH. Heterogeneous oxidation kinetics of organic biomass burning aerosol surrogates by O_3, NO_2, N_2O_5, and NO_3. Phys Chem Chem Phys. 2011;13(47): 21050–21062.

[76] Iannone R, Xiao S, Bertram AK. Potentially important nighttime heterogeneous chemistry: NO_3 with aldehydes and N_2O_5 with alcohols. Phys Chem Chem Phys. 2011;13(21): 10214–10223.

[77] Gross S, Iannone R, Xiao S, Bertram AK. Reactive uptake studies of NO_3 and N_2O_5 on alkenoic acid, alkanoate, and polyalcohol substrates to probe nighttime aerosol chemistry. Phys Chem Chem Phys. 2009;11(36):7792–7803.

[78] Gross S, Bertram AK. Reactive uptake of NO_3, N_2O_5, NO_2, HNO_3, and O_3 on three types of polycyclic aromatic hydrocarbon surfaces. J Phys Chem A. 2008;112(14):3104–3113.

[79] Ryder OS, Campbell NR, Morris H, Forestieri SD, Ruppel MJ, Cappa CD, et al. Role of organic coatings in regulating N_2O_5 reactive uptake to sea spray aerosol. J Phys Chem A. 2015;119 (48):11683–11692.

[80] Gaston CJ, Thornton JA, Ng NL. Reactive uptake of N_2O_5 to internally mixed inorganic and organic particles: The role of organic carbon oxidation state and inferred organic phase separations. Atmos Chem Phys. 2014;14:5693–5707.

[81] Grassian VH. Heterogeneous uptake and reaction of nitrogen oxides and volatile organic compounds on the surface of atmospheric particles including oxides, carbonates, soot and mineral dust: Implications for the chemical balance of the troposphere. Int Rev Phys Chem. 2001;20(3):467–548.

[82] Tang M, Huang X, Lu K, Ge M, Li Y, Cheng P, et al. Heterogeneous reactions of mineral dust aerosol: Implications for tropospheric oxidation capacity. Atmos Chem Phys. 2017;17: 11727–11777.

[83] Donaldson MA, Bish DL, Raff JD. Soil surface acidity plays a determining role in the atmospheric-terrestrial exchange of nitrous acid. Proc Natl Acad Sci USA. 2014;111(52): 18472–18477.

[84] Starokozhev E, Sieg K, Fries E, Puttmann W. Investigation of partitioning mechanism for volatile organic compounds in a multiphase system. Chemosphere. 2011;82:1482–1488.

[85] Zeineddine MN, Romanias MN, Gaudion V, Riffault V, Thevenet F. Heterogeneous interaction of isoprene with natural Gobi dust. ACS Earth Space Chem. 2017;1:236–243.

[86] Ji Y, Wang H, Li G, An T. Theoretical investigation on the role of mineral dust aerosol in atmospheric reaction: A case of the heterogeneous reaction of formaldehyde with NO_2 onto SiO_2 dust surface. Atmos Environ. 2015;103:207–214.

[87]   Wang X, Sun J, Han D, Bao L, Mei Q, Wei B, et al. Gaseous and heterogeneous reactions of low-molecular-weight (LMW) unsaturated ketones with O_3: Mechanisms, kinetics, and effects of mineral dust in tropospheric chemical processes. Chem Eng J. 2020;395(125083):1–11.

[88]   Banfield JF, Zhang H. Nanoparticles in the Environment. In: Banfield JF, Navrotsky A, editors. Reviews in Mineralogy and Geochemistry. vol. 44, Washington DC: Mineralogical Society of America; 2001; 1–58.

[89]   Cornell RM, Schwertmann U. The Iron Oxides: Structure, Properties, Reactions, Occurrences, and Uses. 2nd ed, Weinheim: Wiley-VCH; 2003.

[90]   Pereira MC, Oliveira LCA, Murad E. Iron oxide catalysts: Fenton and Fenton-like reactions – a review. Clay Miner. 2012;47:285–302.

[91]   Al-Abadleh HA. Aging of atmospheric aerosols and the role of iron in catalyzing brown carbon formation. Environ Sci: Atmos. 2021.doi:https://doi.org/10.1039/D1EA00038A.

[92]   Al-Abadleh HA. A review on the bulk and surface chemistry of iron in atmospherically-relevant systems containing humic like substances. RSC Adv. 2015;5(57):45785–45811.

[93]   Tofan-Lazar J, Situm A, Al-Abadleh HA. DRIFTS studies on the role of surface water in stabilizing catechol-iron(III) complexes at the gas/solid interface. J Phys Chem A. 2013;117: 10368–10380.

[94]   Gulley-Stahl H, Hogan IIPA, Schmidt WL, Wall SJ, Buhrlage A, Bullen HA. Surface complexation of catechol to metal oxides: An ATR-FTIR adsorption and dissolution study. Environ Sci Technol. 2010;44:4116–4121.

[95]   Yang Y, Yan W, Jing C. Dynamic adsorption of catechol at the goethite/aqueous solution interface: A molecular-scale study. Langmuir. 2012;28:14588–14597.

[96]   Walker RA. Faster chemistry at surfaces. Nature Chem. 2021;13:296–305.

[97]   Zhong J, Kumar M, Francisco JS, Zeng XC. Insight into chemistry on cloud/aerosol water surfaces. Acc Chem Res. 2018;51(5):1229–1237.

[98]   Kusaka R, Nihonyanagi S, Tahera T. The photochemical reaction of phenol becomes ultrafast at the air–water interface. Nature Chem. 2021;13:306–311.

[99]   Nissenson P, Dabdub D, Das R, Maurino V, Minero C, Vione D. Evidence of the water-cage effect on the photolysis of NO_3^-- and $FeOH^{2+}$. Implications of This Effect and of H_2O_2 Surface Accumulation on Photochemistry at the Air/water Interface of Atmospheric Droplets Atmos Environ. 2010;44:4859–4866.

[100] Nissenson P, Knox CJH, Finlayson-Pitts BJ, Phillips LF, Donald D. Enhanced photolysis in aerosols: Evidence for important surface effects. Phys Chem Chem Phys. 2006;8:4700–4710.

[101] Rossignol S, Tinel L, Bianco A, Passananti M, Brigante M, Donaldson DJ, et al. atmospheric photochemistry at a fatty acid-coated air-water interface. Science. 2016;353(6300):699–702.

[102] Vaida V. Atmospheric radical chemistry revisited. Science. 2016;353(6300):650.

[103] Costa MTC, Anglada JM, Francisco JS, Ruiz-Lopez MF. Reactivity of volatile organic compounds at the surface of a water droplet. J Am Chem Soc. 2012;134(28):11821–11827.

[104] Zhong J, Kumar M, Anglada JM, Martins-Costa MTC, Ruiz-Lopez MF, Zeng XC, et al. Atmospheric spectroscopy and photochemistry at environmental water interfaces. Annu Rev Phys Chem. 2019;70:45–69.

[105] Mekic M, Wang Y, Loisel G, Vione D, Gligorovski S. Ionic strength effect alters the heterogeneous ozone oxidation of methoxyphenols in going from cloud droplets to aerosol deliquescent particles. Environ SciTechnol. 2020;54(20):12898–12907.

[106] Rifkha Kameel F, Riboni F, Hoffmann MR, Enami S, Colussi AJ. Fenton oxidation of gaseous isoprene on aqueous surfaces. J Phys Chem C. 2014;118:29151–29158.

[107] Knox KJ. Light-induced Processes in Optically-tweezed Aerosol Droplets. Heidelberg: Springer; 2011.

[108] Sullivan RC, Boyer-Chelmo H, Gorkowski K, Beydoun H. Aerosol optical tweezers elucidate the chemistry, acidity, phase separations, and morphology of atmospheric microdroplets. Acc Chem Res. 2020;53(11):2498–2509.

[109] Chang YP, Wu SJ, Lin MS, Chiang CY, Huang GG. Ionic-strength and pH dependent reactivities of ascorbic acid toward ozone in aqueous micro-droplets studied using aerosol optical tweezers. Phys Chem Chem Phys. 2021;23:10108–10117.

[110] Zaera F. Probing liquid/solid interfaces at the molecular level. Chem Rev. 2012;112: 2920–2986.

[111] Campbell CT, Sauer J. Introduction: Surface chemistry of oxides. Chem Rev. 2013;113: 3859–3862.

[112] Stumm W, Morgan JJ. Aquatic Chemistry. 3rd edn, New York: John Wiley & Sons, Inc.; 1996.

[113] Sparks DL. Environmental Soil Chemistry. 1st edn, San Diego: Academic Press; 1995.

[114] Hochella MF Jr, Lower SK, Maurice PA, Penn RL, Sahai N, Sparks DL, et al. Nanominerals, mineral nanoparticles, and earth systems. Science. 2008;319:1631–1635.

[115] Al-Abadleh HA, Grassian VH. Oxide surfaces as environmental interfaces. Surf Sci Rep. 2003;52:63–162.

[116] Al-Abadleh HA, Al-Hosney HA, Grassian VH. Oxide and carbonate surfaces as environmental interfaces: Importance of water in surface composition and surface reactivity. J Molec Catal A. 2005;228:47–54.

[117] Brown GE Jr. How minerals react with water. Science. 2001;294(5540):67–69.

[118] Kosmulski M. Surface Charging and Points of Zero Charge. Boca Raton, FL: CRC Press; 2009.

[119] Kosmulski M. Isoelectric points and points of zero charge of metal (hydr)oxides: 50 years after Parks' review. Adv Coll Inter Sci. 2016;238:1–61.

[120] Duval Y, Mielczarski JA, Pokrovsky OS, Mielczarski E, Erhrhardt JJ. Evidence for the existence of three types of species at the quartz-aqueous solution interface at pH 0-10: XPS surface group quantification and surface complexation modeling. J Phys Chem B. 2002;106: 2937–2945.

[121] Azam MS, Weerman CN, Gibbs-Davis JM. Specific cation effects on the bimodal acid–base behavior of the silica/water interface. J Phys Chem Lett. 2012;3:1269–1274.

[122] Gibbs-Davis JM, Kruk JJ, Konek C, Scheidt KA, Geiger FM. Jammed acid-base reactions at interfaces. J Am Chem Soc. 2008;130:15444–15447.

[123] Brown GE Jr., Parks GA, O'Day PA. Sorption at mineral-water interfaces: Macroscopic and microscopic perspectives. In: Vaughan DJ, Pattrick RAD, editors. Mineral Surfaces. New York: Chapman & Hall; 1995. p. 129–183.

[124] Brown GE, Parks GA. Sorption of trace elements on mineral surfaces: Modern perspectives from spectroscopic studies, and comments on sorption in the marine environment. Int Geo Rev. 2001;43:963–1073.

[125] Philippe A, Schaumann GE. Interactions of dissolved organic matter with natural and engineered inorganic colloids: A review. Environ Sci Technol. 2014;48:8946–8962.

[126] Davis JA, Kent DB. Surface Complexation Modeling in Aqueous Geochemistry. In: Hochella MF Jr, Af W, editors. Reviews in Mineralogy: Mineral-Water Interface Geochemistry. vol. 23, Washington D.C.: The Mineralogical Society of America; 1990.

[127] Wh VR, Hiemstra T. Chapter 8 – The CD-MUSIC model as a framework for interpreting ion adsorption on metal (hydr) oxide surfaces. In: Lützenkirchen J, editor. Surface Complexation Modelling. The Netherlands: Elsevier; 2006. p. 251–268.

[128] Benner R. Loose ligands and available iron in the ocean. Proc Natl Acad Sci USA. 2011;108(3): 893–894.

[129] Boyd PW, Ellwood MJ. The biogeochemical cycle of iron in the ocean. Nature Geosci. 2010;3: 675–682.

[130] Schroth AW, Crusius J, Sholkovitz ER, Bostick BC. Iron solubility driven by speciation in dust sources to the ocean. Nat Geosci. 2009;2:337–340.

[131] Meskhidze N, Volker C, Al-Abadleh HA, Barbeau K, Bressac M, Buck C, et al. Perspective on identifying and characterizing the processes controlling iron speciation and residence time at the atmosphere-ocean interface. Mar Chem. 2019;217(103704):1–16.

[132] Wurl O, Ekau W, Wm L, Cj Z. Sea surface microlayer in a changing ocean – A perspective. Elem Sci Anth. 2017;5(31):1–11.

[133] Chernyshova IV, Ponnurangam S, Somasundaran P. Linking interfacial chemistry of CO_2 to surface structures of hydrated metal oxide nanoparticles: Hematite. Phys Chem Chem Phys. 2013;15:6953–6964.

[134] Rubasinghege G, Lentz RW, Scherer MM, Grassian VH. Simulated atmospheric processing of iron oxyhydroxide minerals at low pH: Roles of particle size and acid anion in iron dissolution. Proc Natl Acad Sci USA. 2010;107(15):6628–6633.

[135] Lanzl CA, Baltrusaitis J, Cwiertny DM. Dissolution of hematite nanoparticle aggregates: influence of primary particle size, dissolution mechanism, and solution pH. Langmuir. 2012;28:15797–15808.

[136] Chen H, Grassian VH. Iron dissolution of dust source materials during simulated acidic processing: The effect of sulfuric, acetic, and oxalic acids. Environ Sci Technol. 2013;47(18): 10312–10321.

[137] Chen H, Laskin A, Baltrusaitis J, Gorski CA, Scherer MM, Grassian VH. Coal fly ash as a source of iron in atmospheric dust. Environ Sci Technol. 2012;46:2112–2120.

[138] Gonzalez MC, Roman ES. Environmental Photochemistry in Heterogeneous Media. Hdb Environmental Chemistry. Vol. 2, Berlin: Springer-Verlag; 2005, p. 49–75.

[139] George C, D'Anna B, Herrmann H, Weller C, Vaida V, Donaldson DJ, et al. Emerging Areas in Atmospheric Photochemistry. In: McNeill VF, Ariya PA, editors. Atmospheric and Aerosol Chemistry. vol. 339, Heidelberg: Springer; 2012. p. 1–54.

[140] Link N, Removski N, Yun J, Fleming L, Nizkorodov SA, Bertram AK, et al. Dust-catalyzed oxidative polymerization of catechol under acidic conditions and its impacts on ice nucleation efficiency and optical properties. ACS Earth Space Chem. 2020;4(7):1127–1139.

[141] Sparks DL. Metal and Oxyanion Sorption on Naturally Occurring Oxide and Clay Mineral Surfaces. In: Grassian VH, editor. Environmental Catalysis. Boca Raton: Taylor & Francis Group; 2005. p. 3–36.

[142] Kubicki JD, Kwon KD, Paul KW, Sparks DL. Surface complex structures modelled with quantum chemical calculations: Carbonate, phosphate, sulphate, arsenate and arsenite. Europ J Soil Sci. 2007;58:932–944.

[143] Violante A. Elucidating Mechanisms of Competitive Sorption at the Mineral/Water Interface. In: Sparks DL, editor. Advances in Agronomy. Vol. 118, San Diego: Elsevier Academic Press Inc; 2013. p. 111–176.

[144] Cygan RT, Kubicki JD. editors. Molecular Modeling Theory: Applications in the Geosciences. Washington DC: Mineralogical Society of America; 2001.

[145] Mudunkotuwa IA, Grassian VH. Biological and environmental media control oxide nanoparticle surface composition: The roles of biological components (proteins and amino acids), inorganic oxyanions and humic acid. Environ Sci Nano. 2015;2:429–439.

[146] Kerisit S, Ilton ES, Parker SC. Molecular dynamics simulations of electrolyte solutions at the (100) goethite surface. J Physl Chem B. 2006;110:20491–20501.

[147] J-f B, Shchukarev A. X-ray photoelectron spectroscopy of fast-frozen hematite colloids in aqueous solutions. 2. Tracing the relationship between surface charge and electrolyte adsorption. J Phys Chem C. 2010;114(6):2613–2616.

[148] Shimizu K, Shchukarev A, Kozin PA, Boily J-F. X-ray photoelectron spectroscopy of fast-frozen hematite colloids in aqueous solutions. 4. Coexistence of alkali metal (Na^+, K^+, Rb^+, Cs^+) and chloride ions. Surf Sci. 2012;606(13-14):1005–1009.

[149] Shimizu K, Shchukarev A, Kozin PA, Boily J-F. X-ray photoelectron spectroscopy of fast-frozen hematite colloids in aqueous solutions. 5. Halide Ion (F^-, Cl^-, Br^-, I^-) adsorption. Langmuir. 2013;29(8):2623–2630.

[150] Situm A, Rahman MA, Allen N, Kabengi NJ, Al-Abadleh HA. ATR-FTIR and flow microcalorimetry studies on the initial binding kinetics of arsenicals at the organic-hematite interface. J Phys Chem A. 2017;121:5569–5579.

[151] Situm A, Rahman MA, Goldberg S, Al-Abadleh HA. Spectral characterization and surface complexation modeling of organics on hematite nanoparticles: Role of electrolytes in the binding mechanism. Environ Sci Nano. 2016;3:910–926.

[152] Paul KW, Kubicki JD, Sparks DL. Quantum Chemical calculations of sulfate adsorption at the Al- and Fe-(hydr)oxide-H_2O interface-estimation of Gibbs fee energies. Environ Sci Technol. 2006;40:7717–7724.

[153] Eggleston CM, Hug SJ, Stumm W, Sulzberger B, Afonso MDS. Surface complexation of sulfate by hematite surfaces: FTIR and STM observations. Geochem Cosmochim Acta. 1998;62(4): 585–593.

[154] Hug SJ. In situ Fourier transform infrared measurements of sulfate adsorption on hematite in aqueous solutions. J Coll Inter Sci. 1997;188:415–422.

[155] Han J, Kim M, Ro H-M. Factors modifying the structural configuration of oxyanions and organic acids adsorbed on iron (hydr)oxides in soils. A Review Environ Chem Lett. 2020;18:631–662.

[156] Pincus LN, Rudel HE, Petrovic PV, Gupta S, Westerhoff P, Muhich CL, et al. Exploring the mechanisms of selectivity for environmentally significant oxo-anion removal during water treatment: A review of common competing oxo-anions and tools for quantifying selective adsorption. Environ Sci Technol. 2020;54:9769–9790.

[157] Baltrusaitis J, Schuttlefield J, Jensen JH, Grassian VH. FTIR spectroscopy combined with quantum chemical calculations to investigate adsorbed nitrate on aluminium oxide surfaces in the presence and absence of co-adsorbed water. Phys Chem Chem Phys. 2007;9: 4970–4980.

[158] Wang X, Wang Z, Peak D, Tang Y, Feng X, Zhu M. Quantification of coexisting inner- and outer-sphere complexation of sulfate on hematite surfaces. ACS Earth Space Chem. 2018;2:387–398.

[159] Baltrusaitis J, Grassian VH. Carbonic acid formation from reaction of carbon dioxide and water coordinated to $Al(OH)_3$: A quantum chemical study. J Phys Chem A. 2010;114: 2350–2356.

[160] Baltrusaitis J, Schuttlefield J, Zeitler E, Jensen JH, Grassian VH. Surface reactions of carbon dioxide at the adsorbed water-oxide interface. J Phys Chem C. 2007;111:14870–14880.

[161] Baltrusaitis J, Grassian VH. Surface reactions of carbon dioxide at the adsorbed water-iron oxide interface. J Phys Chem B Lett. 2005;109:12227–12230.

[162] Plata JJ, Collico V, Marquez AM, Sanz JF. Understanding acetaldehyde thermal chemistry on the TiO2 (110) rutile surface: From adsorption to reactivity. J Phys Chem C. 2011;115(6): 2819–2825.

[163] Subdiaga E, Harir M, Orsetti S, Hertkorn N, Schmitt-Kopplin P, Haderlein SB. Preferential sorption of tannins at aluminum oxide affects the electron exchange capacities of dissolved and sorbed humic acid fractions. Environ Sci Technol. 2020;54(3):1837–1847.

[164] Shen Z, Zhang Z, Li T, Yao Q, Zhang T, Chen W. Facet-dependent adsorption and fractionation of natural organic matter on crystalline metal oxide nanoparticles. Environ Sci Technol. 2020;54(14):8622–8631.

[165] Johnson SB, Yoon TH, Kocar BD, Brown GE Jr. Adsorption of organic matter at mineral/water interfaces. Outer sphere adsorption of maleate and implications for dissolution processes. Langmuir. 2004;20:4996–5006.

[166] Bondietti G, Sinniger J, Stumm W. III. The reactivity of Fe(III) (hydr)oxides: Effects of ligands in inhibiting the dissolution. Colloids Surf A. 1993;79:157–167.

[167] Colarieti ML, Toscano G, Ardi MR, Greco JG. Abiotic oxidation of catechol by soil metal oxides. J Haz Mater B. 2006;134:161–168.

[168] Larson RA, Hufnal JM Jr. Oxidative polymerization of dissolved phenols by soluble and insoluble inorganic species. Limnol Oceanogr. 1980;25(3):505–512.

[169] Slikboer S, Grandy L, Blair SL, Nizkorodov SA, Smith RW, Al-Abadleh HA. Formation of light absorbing soluble secondary organics and insoluble polymeric particles from the dark reaction of catechol and guaiacol with Fe(III). Environ Sci Technol. 2015;49(13):7793–7801.

[170] Deguillaume L, Leriche M, Desboeufs K, Mailhot G, George C, Chaumerliac N. Transition metals in atmospheric liquid phases: sources, reactivity, and sensitive parameters. Chem Rev. 2005;105(9):3388–3431.

[171] Win MS, Tian Z, Zhao H, Xiao K, Peng J, Shang Y, et al. Atmospheric HULIS and its ability to mediate the reactive oxygen species (ROS): A review. J Environ Sci. 2018;71:13–31.

[172] Schneider J, Matsuoka M, Takeuchi M, Zhang J, Horiuchi YU, Anpo M, et al. Understanding TiO2 photocatalysis: Mechanisms and materials. Chem Rev. 2014;114:9919–9986.

[173] Chen H, Nanayakkara CE, Grassian VH. Titanium dioxide photocatalysis in atmospheric chemistry. Chem Rev. 2012;112:5919–5948.

[174] Vione D, Maurino V, Minero C, Pelizzetti E, Harrison MAJ, Olariu R-I, et al. Photochemical reactions in the tropospheric aqueous phase and on particulate matter. Chem Soc Rev. 2006;35:441–453.

[175] Cwiertny DM, Young MA, Grassian VH. Chemistry and photochemistry of mineral dust aerosol. Annu Rev Phys Chem. 2008;59:27–51.

[176] Chen H, Navea JG, Young MA, Grassian VH. Heterogeneous photochemistry of trace atmospheric gases with components of mineral dust aerosol. J Phys Chem A. 2011;115: 490–499.

[177] Ndour M, D'Anna B, George C, Ka O, Balkanski Y, Kleffmann J, et al. Photoenhanced uptake of NO2 on mineral dust: Laboratory experiments and model simulations. Geophys Res Lett. 2008;35:L05812.

[178] Xu B, Zhu T, Tang X, Shang J. Heterogeneous reaction of formaldehyde on the surface of TiO_2 particles. Sci China Chem. 2010;53(12):2644–2651.

[179] Sassine M, Burel L, D'Anna B, George C. Kinetics of the tropospheric formaldehyde loss onto mineral dust and urban surfaces. Atmos Environ. 2010;44:5468–5475.

[180] Styler SA, Donaldson DJ. Photooxidation of atmospheric alcohols on laboratory proxies for mineral dust. Environ Sci Technol. 2011;45:10004–10012.

[181] Alvarez EG, Wortham H, Strekowski RS, Zetzsch C, Gligorovski S. Atmospheric photosensitized heterogeneous and multiphase reactions: From outdoors to indoors. Environ Sci Technol. 2012;46(4):1955–1963.

[182] Bourikas K, Kordulis C, Lycourghiotis A. Titanium dioxide (Anatase and Rutile): Surface chemistry, liquid–solid interface chemistry, and scientific synthesis of supported catalysts. Chem Rev. 2014;114:9754–9823.

[183] Chu B, Wang Y, Yang W, Ma J, Ma Q, Zhang P, et al. Effects of NO_2 and C_3H_6 on the heterogeneous oxidation of SO_2 on TiO_2 in the presence or absence of UV–Vis irradiation. Atmos Chem Phys. 2019;19:14777–14790.

[184] Fujishima A, Saeki M, Kimizuka N, Taniguchi M, Suga S. Photoemission satellites and electronic structure of Fe2O3. Phys Rev B. 1986;34:7318–7333.

[185] Zhang H, Bayne M, Fernando S, Legg B, Zhu M, Penn RL, et al. Size-dependent bandgap of nanogoethite. J Phys Chem C. 2011;115:17704–17710.

[186] Martinez AI, Garcia-Lobato MA, Perry DL. Study of the Properties of Iron Oxide Nanostructures. In: Barranon A, editor. Research in Nanotechnology Developments. New York: Nova Science Publishers, Inc.; 2009. p. 183–194.

[187] Rubasinghege G, Elzey S, Baltrusaitis J, Jayaweera PM, Grassian VH. Reactions on atmospheric dust particles: Surface photochemistry and size-dependent nanoscale redox chemistry. J Phys Chem Lett. 2010;1:1729–1737.

[188] Kim K, Choi W, Hoffmann MR, Yoon HI, Park BK. Photoreductive dissolution of iron oxides trapped in ice and its environmental implications. Environ Sci Technol. 2010;44:4142–4148.

[189] Leow WR, Ng WKH, Peng T, Liu X, Li B, Shi W, et al. Al_2O_3 surface complexation for photocatalytic organic transformations. J Am Chem Soc. 2017;139:269–276.

[190] Chu B, Liu T, Zhang X, Liu Y, Ma Q, Ma J, et al. Secondary aerosol formation and oxidation capacity in photooxidation in the presence of Al_2O_3 seed particles and SO_2. Sci China Chem. 2015;58(9):1426–1434.

[191] Zhang X, Chu B, Li J, Zhang C. Effects of seed particles Al_2O_3, $Al_2(SO_4)_3$ and H_2SO_4 on secondary organic aerosol. Front Environ Sci Eng. 2017;11(5), doi:https://doi.org/10.1007/s11783-017-0936-4.

[192] Chu B, Zhang X, Liu Y, He H, Sun Y, Jiang J, et al. Synergetic formation of secondary inorganic and organic aerosol: Effect of SO_2 and NH_3 on particle formation and growth. Atmos Chem Phys. 2016;16:14219–14230.

[193] Richard C, Canonica S. Aquatic phototransformation of organic contaminants induced by coloured dissolved organic matter. In: Hdb Environmental Chemistry. Vol. 2, Berlin: Springer-Verlag; 2005. p. 299–323.

[194] Young MA. Environmental Photochemistry in Surface Waters. In: Lehr J, Kelley J, editors. Water Encyclopedia: Oceanography; Meteorology; Physics and Chemistry; Water Law; and Water History, Art, and Culture. New York: Wiley & Sons; 2005. p. 529–540.

[195] Chan MN, Chan CK. Hygroscopic properties of two model humic-like substances and their mixtures with inorganics of atmospheric importance. Environ Sci Techn. 2003;37:5109–5155.

[196] Dinar E, Taraniuk I, Graber ER, Anttila T, Mentel TF, Rudich Y. Hygroscopic growth of atmospheric and model humic-like substances. J Geophys Res-Atmos. 2007;112:D05211, doi:10.1029/2006JD007442.

[197] Chan MN, Kreidenweis SM, Chan CK. Measurements of the hygroscopic and deliquescence properties of organic compounds of different solubilities in water and their relationship with cloud condensation nuclei activities. Environ Sci Technol. 2008;42(10):3602–3608.

[198] Zelenya V, Huthwelker T, Krepelova A, Rudich Y, Ammann M. Humidity driven nanoscale chemical separation in complex organic matter. Environ Chem. 2011;8(4):450–460.

[199] Cowen S, Al-Abadleh HA. DRIFTS studies on the photodegradation of tannic acid as a model for HULIS in atmospheric aerosols. Phys Chem Chem Phys. 2009;11(3):7838–7847.

[200] Stemmler K, Ammann M, Donders C, Kleffmann J, George C. Photosensitized reduction of nitrogen dioxide on humic acid as a source of HONO. Nature. 2006;440:195–198.

[201] Sosedova Y, Rouviere A, Bartels-Rausch T, Ammann M. UVA/Vis-induced nitrous acid formation on polyphenolic films exposed to gaseous NO_2. Photochem Photobiol Sci. 2011;10 (10):1680–1690.
[202] Ofner J, Kruger H-U, Grothe H, Schmitt-Kopplin P, Whitmore K, Zetzsch C. Physico-chemical characterization of SOA derived from catechol and guaiacol – A model substance for the aromatic fraction of atmospheric HULIS. Atmos Chem Phys. 2011;11:1–15.

Chapter 5
Bulk aqueous phase chemistry relevant to cloud droplets

The bulk aqueous phase reactions of relevance to cloud droplets have been the subject of thematic and relatively recent reviews by Tilgner et al. [1], Herrmann et al., Deguillaume et al. [2], Gligorovski et al. [3], and Bianco et al. [4]. These reactions are promoted or affected by gas phase partitioning, acidity of the aqueous phase, dissolution products of insoluble cores such as mineral dust. These reactions can result in forming secondary organic and inorganic species. In multicomponent aqueous systems, synergetic reactions take place due to in situ reactive radical formation or absorption of light. Understanding the underlying mechanisms of these processes is needed for modeling chemical processing in cloud droplets and wet aerosol particles [5]. The following sections provide an overview of selected dark and photochemical reactions with examples describing chemistry in bulk water and microliter-sized droplets.

5.1 Dark reactions

5.1.1 Oxidized and reduced nitrogen

The nitrate radical is the main nighttime oxidant. Using laser photolysis, de Sémainville et al. [6] studied the reactivity of nitrate radical toward four dicarboxylic acids: malonic acid, mesoxalic acid, oxalic acid, and succinic acid under various pH and temperature values. These reactions occur either by hydrogen abstraction at the C–H bond or charge exchange reactions with the carboxylate groups. The effect of speciation on reactivity was studied by varying the pH and showed that the first mechanism proceeds at all pH values while the second one proceeds for pH values above the first pK_a. Hence, under mildly acidic conditions relevant to atmospheric particles, reactions with the nitrate radical compete with the oxidation by hydroxyl radical. Temperature-dependent studies of the rate constants allowed for the determination of the activation energy barrier, which were diacid-specific.

Also, aqueous phase reactions between atmospheric carbonyls and ammonia/amines were reported to form reduced-nitrogen organic compounds regarded as important brown carbon (BrC) chromophores [7]. For example, the reaction of methylglyoxal with $(NH_4)_2SO_4$ was found to form 30 BrC nitrogen-containing and light-absorbing ($\lambda_{max} > 272$–366 nm) chromophores with 6–12 carbon atoms [8]. In the presence of

https://doi.org/10.1515/9781501519376-005

aqueous ammonia, NH_3, in equilibrium with NH_4^+, imidazole and its derivatives were formed from the reaction with glyoxal [9].

5.1.2 Transition metal-driven reactions

Transition metals of atmospheric relevance include iron (Fe), copper (Cu), and manganese (Mn) with concentrations ranging from 10^{-7} to 10^{-6} M in cloud water with Fe species present at ~10× that of Cu and Mn [2, 10]. Based on kinetic and speciation

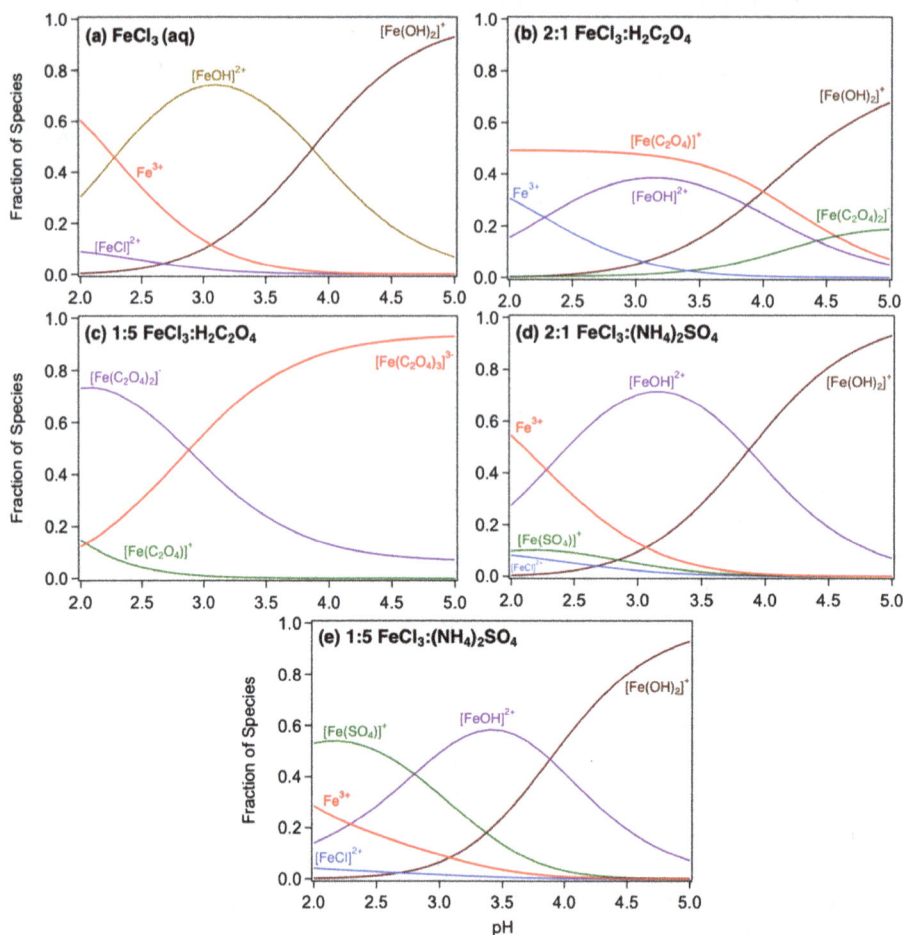

Figure 5.1: Speciation curves of iron chloride, oxalate, and sulfate with variable molar ratios. Ratios in headings are mol:mol. The curves were generated using equilibrium constants for the acid dissociation and complexation reactions of iron from the database in Visual MINTEQ, v. 3.1. Figure and caption were reproduced with permission from reference [12], © The American Chemical Society, 2019.

modeling conducted on field samples [11], the dominant oxidation states of these metals in fog and cloud water during daytime are Fe(II), Cu(II), and Mn(II) with increasing Cu(I) and Mn(III). During nighttime conditions, Fe(III), Cu(II), and Mn(II) are predominant oxidations states. Of the three metals, the speciation of Fe(III) varies with pH as shown in Figure 5.1 depending on the ligands in solution, and is predominantly $Fe(OH)^{2+}$ between pH 2.5 and 4. Identifying and quantifying aqueous phase species is the first step for understanding the reactivity of systems containing metals and organic/inorganic ligands.

These transition metals react with reactive oxygen species (ROS), sulfur, nitrogen, and chlorine species present in cloud water. The following sections summarize the relatively fastest reactions as per the chemical mechanism model proposed by Deguillaume et al. [2]. These reactions were reported in the absence of organic ligands that strongly complex with transition metals. Section 5.1.2.5 describes complexation reactions and their effect on reactivity of transition metals.

5.1.2.1 Reactivity with ROS

In the case of iron, cycling between oxidation states II and III leads to the consumption and production of ROS [2]. For example, Fenton reaction shown in eq. (5.1) produces hydroxyl radicals [13], which can further react with excess Fe(II) as per eq. (5.2):

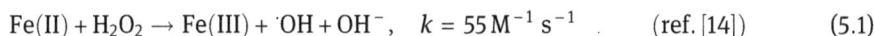

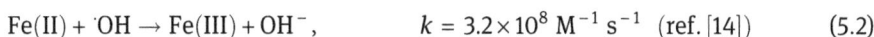

$$Fe(II) + H_2O_2 \rightarrow Fe(III) + \dot{O}H + OH^-, \quad k = 55\,M^{-1}\,s^{-1} \quad \text{(ref. [14])} \quad (5.1)$$

$$Fe(II) + \dot{O}H \rightarrow Fe(III) + OH^-, \quad k = 3.2 \times 10^8\,M^{-1}\,s^{-1} \quad \text{(ref. [14])} \quad (5.2)$$

The rate and intermediates formed in this reaction depend on pH. Highest rates of $\dot{O}H$ production, measured indirectly, were recorded at pH < 3 [13, 15, 16]. Fenton-like reactions take place when Fe(III) or hydroperoxide groups (ROOH) are present instead of H_2O_2 producing alkoxy radicals (RO˙) or alkoxy anions as follows:

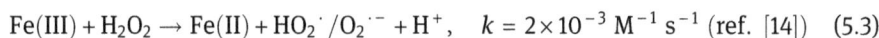

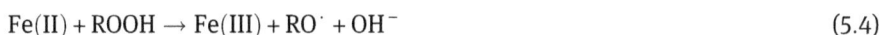

$$Fe(III) + H_2O_2 \rightarrow Fe(II) + HO_2\dot{}/O_2^{\dot{}-} + H^+, \quad k = 2 \times 10^{-3}\,M^{-1}\,s^{-1} \quad \text{(ref. [14])} \quad (5.3)$$

$$Fe(II) + ROOH \rightarrow Fe(III) + RO\dot{} + OH^- \quad (5.4)$$

Rate constants for reaction (5.4) will vary depending on the R group. Using newly activated cloud droplets, Paulson et al. [17] reported a "burst" of $\dot{O}H$ radical formation upon irradiation, which they found to be similar to the photoreaction between peracetic acid and Fe(II) [17]. Fang et al. [18] studied the reactivity of isoprene hydroxy hydroperoxides (ISOPOOH) with Fe(II) in the presence of oxalate. ISOPOOH is formed by the photo-oxidation of isoprene under low NO_x conditions and is a major contributor to secondary organic aerosol (SOA) formation. For reaction (5.4), they used a rate constant of $4.0 \times 10^4\,M^{-1}\,s^{-1}$ in their kinetic model. Upon reaction with Fe(II), $\dot{O}H$ and organic radicals were observed at room temperature. Iron oxalate complexes formed at low oxalate concentration promoted $\dot{O}H$ formation by

ISOPOOH. However, excess oxalate scavenged ˙OH radicals, thereby lowering its concentration in the aqueous phase. Such chemistry influences the lifetime of organics in the aqueous phase via ROS pathways.

Also, Fe(II) and Fe(III) can react with $HO_2˙/O_2˙^-$ according to reactions (5.5)–(5.7):

$$O_2˙^- + Fe(II) + 2H^+ \rightarrow Fe(III) + H_2O_2, \quad k = 1.0 \times 10^7 \text{ M}^{-1}\text{s}^{-1} \text{ (ref.[10])} \tag{5.5}$$

$$HO_2˙ + Fe(II) + H^+ \rightarrow Fe(III) + H_2O_2, \quad k = 1.2 \times 10^6 \text{ M}^{-1}\text{s}^{-1} \text{ (ref.[10])} \tag{5.6}$$

$$O_2˙^- + FeOH^{2+} \rightarrow FeOH^+ + O_2, \quad k = 1.5 \times 10^8 \text{ M}^{-1}\text{s}^{-1} \text{ (ref.[10])} \tag{5.7}$$

In the case of copper, the following reactions with H_2O_2 and $HO_2˙/O_2˙^-$ were reported:

$$H_2O_2 + Cu(I) \rightarrow Cu(II) + ˙OH + OH^-, \quad k = 7.0 \times 10^3 \text{ M}^{-1}\text{s}^{-1} \text{ (ref. [2])} \tag{5.8}$$

$$HO_2˙ + Cu(II) \rightarrow Cu(I) + O_2 + H^+, \quad k = 1.0 \times 10^8 \text{ M}^{-1}\text{s}^{-1} \text{ (ref. [19])} \tag{5.9}$$

$$HO_2˙ + Cu(I) + H^+ \rightarrow Cu(II) + H_2O_2, \quad k = 2.3 \times 10^9 \text{ M}^{-1}\text{s}^{-1} \text{ (ref. [10])} \tag{5.10}$$

$$O_2˙^- + Cu(II) \rightarrow Cu(I) + O_2, \quad k = 8.0 \times 10^9 \text{ M}^{-1}\text{s}^{-1} \text{ (ref. [10])} \tag{5.11}$$

$$O_2˙^- + Cu(I) + 2H^+ \rightarrow Cu(II) + H_2O_2, \quad k = 9.4 \times 10^9 \text{ M}^{-1}\text{s}^{-1} \text{ (ref. [10])} \tag{5.12}$$

As detailed in reference [2], another important reduction pathway of Fe(III) during night and daytime in cloud water is the direct reaction with Cu(I) as per the following equation:

$$Fe(III)/Fe(OH)^{2+} + Cu(I) \rightarrow Fe(II) + Cu(II) + OH^-, \quad k = 1.3 \times 10^7 \text{ M}^{-1}\text{s}^{-1} \text{ (ref. [2])} \tag{5.13}$$

Since both iron and copper species act as sinks for ROS, if the copper/iron ratio exceeds 1–2%, then the Cu(I)/Cu(II) redox pair becomes the dominant superoxide radical sink [2].

As per the rate constants listed in reference [2], Mn(II) and Mn(III) react with $O_2˙^-$, ˙OH, $HO_2˙$ and H_2O_2:

$$O_2˙^- + Mn(II) \rightarrow MnO_2^+, \quad k = 9.5 \times 10^7 \text{ M}^{-1}\text{s}^{-1} \text{ (ref. [2])} \tag{5.14}$$

$$˙OH + Mn(II) \rightarrow Mn(OH)^{2+}, \quad k = 2.0 \times 10^7 \text{ M}^{-1}\text{s}^{-1} \text{ (ref. [2])} \tag{5.15}$$

$$HO_2˙ + Mn(II) \rightarrow MnO_2^+ + H^+, \quad k = 1.45 \times 10^6 \text{ M}^{-1}\text{s}^{-1} \text{ (ref. [2])} \tag{5.16}$$

$$H_2O_2 + Mn(OH)^{2+} \rightarrow MnO_2^+ + H^+ + H_2O, \quad k = 2.8 \times 10^3 \text{ M}^{-1}\text{s}^{-1} \text{ (ref. [2])} \tag{5.17}$$

5.1.2.2 Reactivity with sulfur

Sulfur dioxide (SO_2) is one of the gases that is efficiently transferred to the aqueous phase forming bisulfite and sulfite ions as per the following reactions:

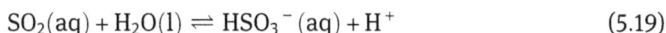

$$SO_2(g) + H_2O(l) \rightleftharpoons SO_2(aq) \tag{5.18}$$

$$SO_2(aq) + H_2O(l) \rightleftharpoons HSO_3^-(aq) + H^+ \tag{5.19}$$

$$HSO_3^-(aq) \rightleftharpoons SO_3^{2-}(aq) + H^+ \tag{5.20}$$

The role of transition metals in catalyzing the oxidation of S(IV) in SO_2(aq), HSO_3^- and SO_3^{2-}, to S(VI) in H_2SO_4 received extensive attention given its role in acid rain formation [2, 10, 20]. Harris et al. [21] found that the dominant in-cloud oxidation pathway of SO_2 oxidation is catalyzed by transition metal ions (namely Fe(III) and Mn(II)) over the H_2O_2 oxidation pathway at pH 4 relevant to cloud water. Scheme 5.1 shows the synergy between Fe(III) and Mn(II) in oxidizing HSO_3^- [22]:

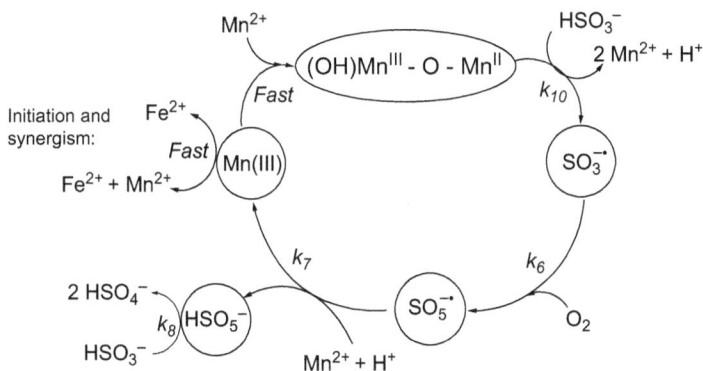

Scheme 5.1: Catalytic cycle for the manganese-catalyzed autoxidation of hydrogen sulfite at pH 2.4, including iron/manganese synergism and chain initiation by trace concentrations of iron(III). Notation for rate constants is according to the equations in the text of reference [22] and values of rate constants are $k_6 = 2.5 \times 10^9$, 1.1×10^9 $M^{-1} s^{-1}$, $k_7 = 10^8$ $M^{-1} s^{-1}$, $k_8 = 4.3 \times 10^4$ $M^{-1} s^{-1}$, and $k_{10} = 1.5 \times 10^6$, 1.6×10^6 $M^{-1} s^{-1}$. Mn(III) denotes mixed hydroxo aqua complexes of manganese(III). Scheme and caption were reproduced with permission from reference [22], © The American Chemical Society, 1998.

The modeling results reported by Harris et al. [21] had relatively high Ni(II), Zn(II), and Cu(II) concentrations but displayed low transition metal ion (TMI)-catalyzed oxidation, which suggest that these metals are less active in the oxidation of S(IV).

Organosulfates (R-OSO_3H) are ubiquitous constituents of atmospheric aerosol particles attracting attention to their sources and different pathways that lead to their formation [23]. Huang et al. [24] reported the formation of C2–C4 organosulfur compounds from the aqueous reaction of isoprene oxidation products such as methacrolein and methyl vinyl ketone with sulfite/bisulfite in the presence of Fe(III) and Mn(II). Oligomers of the organosulfur compounds with more than eight carbon atoms were observed in the case of Fe(III) only. The reason behind this observation was speculated to be due to complex formation between Mn(II) and organic radicals that inhibited the chain propagation. The same group also reported on the formation of aromatic organosulfates and sulfonates in the Fe(III)-driven aqueous phase reactions of a series of aromatics (e.g., benzoic acid, toluene, and phenol) with S(IV) [25]. The authors reported that the aromatic ring of parent aromatics is retained in these reactions, where the mechanism highlights the role of the combination of organic radical intermediates with sulfoxy radicals $SO_3^{-\cdot}$ and $SO_4^{-\cdot}$.

5.1.2.3 Reactivity with nitrogen oxides

The oxidation state of nitrogen in nitrogen oxides of relevance to atmospheric chemistry spans a wide range from +6 in NO_3 radicals to +5 in HNO_3 and NO_3^- to +4 in NO_2 to +3 in HONO and NO_2^-. The following redox reactions between nitrogen oxides and iron were reported in the review by Deguillaume et al. [2]:

$$NO_3 + Fe(II) \rightarrow Fe(III) + NO_3^-, \quad k = 8.0 \times 10^6 \ M^{-1} s^{-1} \quad (5.21)$$

$$NO_2 + Fe(II) \rightarrow Fe(III) + NO_2^-, \quad k = 3.1 \times 10^5 \ M^{-1} s^{-1} \quad (5.22)$$

$$HONO + FeO^{2+} \rightarrow Fe(III) + NO_2 + OH^-, \quad k = 1.1 \times 10^4 \ M^{-1} s^{-1} \quad (5.23)$$

$$NO_2^- + FeO^{2+} + H^+ \rightarrow Fe(III) + NO_2 + OH^-, \quad k = 1.0 \times 10^5 \ M^{-1} s^{-1} \quad (5.24)$$

Also, a reaction between NO_3^- and Fe(II) was also reported as per the following equation [26, 27]:

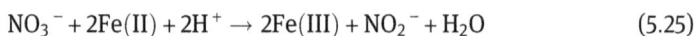

$$NO_3^- + 2Fe(II) + 2H^+ \rightarrow 2Fe(III) + NO_2^- + H_2O \quad (5.25)$$

The rate of reaction (5.25) was found to be fast under neutral to basic pH in aqueous solutions. As noted by Cwiertny et al. [28], nitrate reduction by Fe(II)-containing surfaces was also observed under acidic conditions.

5.1.2.4 Reactivity with chlorine species

Transition metals can complex chloride anions and can react with $Cl_2^{-\cdot}$(aq) species. As reviewed by Gen et al. [29], iron can coexist with chloride (Cl^-) in atmospheric fine particles from coastal and noncoastal (i.e., inland) environments due to anthropogenic and combustion sources. Hence, as an anion ligand, aqueous chloride can complex to iron forming iron chloride complexes in aerosol liquid water. As per the speciation curves in Figure 5.1, aqueous solutions containing iron chloride have $FeCl_2^+$ species at lower concentrations than $FeOH_2^+$, $FeSO_4^+$, and $[Fe(C_2O_4)]^+$. As detailed in Section 5.2, the photodissociation of $FeCl_2^+$ species results in the production of reactive chlorine ($^\bullet$Cl) radicals that can carry on further reactions. Deguillaume et al. [2] listed the following reactions of Fe(II) species:

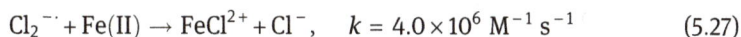

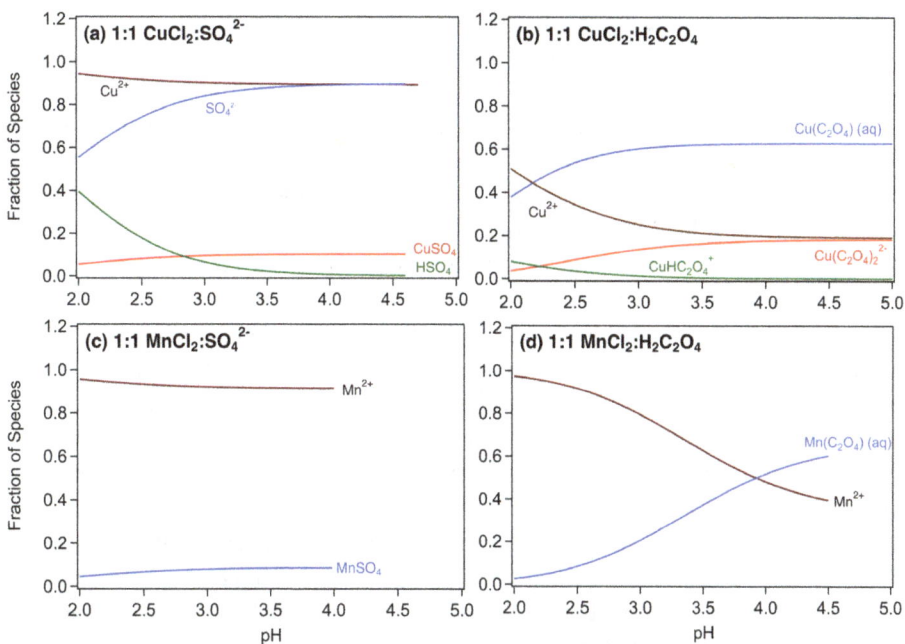

$$Cl_2^{-\bullet} + Fe(II) \rightarrow Fe(III) + 2Cl^-, \quad k = 1.0 \times 10^7 \ M^{-1} \ s^{-1} \tag{5.26}$$

$$Cl_2^{-\bullet} + Fe(II) \rightarrow FeCl^{2+} + Cl^-, \quad k = 4.0 \times 10^6 \ M^{-1} \ s^{-1} \tag{5.27}$$

Figure 5.2: Speciation curves of copper and manganese chloride with sulfate and oxalate as competing ligands. Ratios in headings are mol:mol. The curves were generated using equilibrium constants for the acid dissociation and complexation reactions from the database in Visual MINTEQ, v. 3.1.

Figure 5.2 shows the calculated speciation curves for Cu(II) and Mn(II) chlorides in the presence of sulfate and oxalate as competing ligands. Chloride is a weaker ligand than water, sulfate, and oxalate in the acidic pH range relevant to aerosol particles and cloud droplets. Of these, oxalate is a stronger complexation ligand than sulfate for both metals. Copper oxalate species are dominant in the pH range 2.2–5, whereas manganese oxalate species dominate at pH > 4. Deguillaume et al. [2] listed the following reactions of Cu(I) and Mn(II) species:

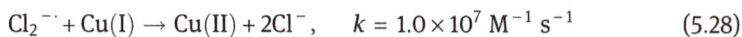

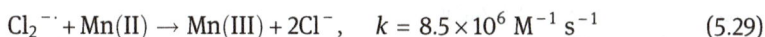

$$Cl_2^{-\bullet} + Cu(I) \rightarrow Cu(II) + 2Cl^-, \quad k = 1.0 \times 10^7 \ M^{-1} \ s^{-1} \tag{5.28}$$

$$Cl_2^{-\bullet} + Mn(II) \rightarrow Mn(III) + 2Cl^-, \quad k = 8.5 \times 10^6 \ M^{-1} \ s^{-1} \tag{5.29}$$

$$Cl_2^{-\cdot} + Mn(II) \rightarrow MnCl_2^+, \qquad k = 2.0 \times 10^5 \ M^{-1} \ s^{-1} \qquad (5.30)$$

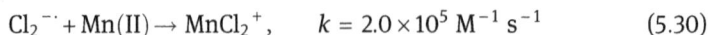

The $MnCl_2^+$ species were reported to decompose to either Mn(II) and $Cl_2^{-\cdot}$ species with a rate constant that equals $3.0 \times 10^5 \ M^{-1} \ s^{-1}$, or Mn(III) and Cl^- with a rate constant that equals $2.1 \times 10^5 \ M^{-1} \ s^{-1}$.

5.1.2.5 Complexation to organic ligands and effects on reactivity

Acidic and phenolic functional groups are strong complexing agents to soluble iron. [30] Visual MINTEQ freeware is a chemical equilibrium computer program with extensive thermodynamic databases that include a large selection of organic acids and inorganic ligands. The database management tool allows organic species to be easily added or deleted. It could be used as a tool to obtain values for the pK_a and complex stability constants (log K) [31]. Also, this program can generate the relative concentration of different species that exist in solution at equilibrium. For example, the supporting information of references [12, 32] lists equilibrium reactions for complex formation of iron species with chloride, sulfate, oxalate, malonate, glutarate, malate, and succinate. Also, the speciation curves shown in Figures 5.1 and 5.2 were generated using Visual MINTEQ. These curves are very useful in quantifying dominant species at a given pH at equilibrium for accurate interpretation of chemical and photochemical reactivities in multicomponent systems. On the other hand, studies on the kinetics of iron complex formation with the organic ligands of atmospheric relevance are sparse. Table 5.1 lists values for the forward complex formation rate constants, k_f, between $FeOH^{2+}$, selected aliphatic carboxylic acids, and catechol. Gen et al. [33] summarized literature values for the formation of iron oxalates and iron malonates under other conditions. Additional kinetic studies are needed using other strong ligands under ionic strength and pH conditions that mimic real aerosol and cloud conditions.

Table 5.1: Literature values for the complex formation of rate constant between $FeOH^{2+}$ and selected ligands.

Reaction	Experimental conditions	Forward rate constant, k_f ($M^{-1} \ s^{-1}$)	References
Citric acid	20 °C, [HClO$_4$] = 0.01–0.05 M, pH = 1–2	50–930	[34]
Oxalic acid	25 °C, 1 M HClO$_4$, "acidic" pH	2×10^4 for FeO_x^+	[35]
Sulfate	25 °C, 1 M NaClO$_4$, pH = 0.7–2.5	1×10^3 for $Fe(SO_4)_2^-$	[36]
DL-Malic acid	25 °C, 1 M NaClO$_4$, pH = 1–2	$95–10^3$	[37]
Catechol	25 °C, 1 M NaClO$_4$, pH = 1–2	3×10^3 for $FeCA^+$	[38]

Depending on the binding affinity of organic ligands to transition metals, ligand exchange reactions occur if they are thermodynamically favorable. For example, ligand exchange between catecholate and iron sulfate complexes is favorable with log $K = 3.6$ according to the following reaction:

$$H(C_6O_2H_4)^- + FeSO_4^+ \rightleftharpoons SO_4^{2-} + Fe(C_6O_2H_4)^+ + H^+ \quad \log K = 3.6 \tag{5.31}$$

Also, the structure of the aliphatic organic ligand will determine if the complex will remain soluble or precipitate out of the solution. Tran et al. [39] showed that aqueous-phase complexation reactions of Fe(III) with fumarate and muconate result in the formation of iron-polyfumarate and iron-polymuconate nanoparticles, whereas reactions with succinate and maleate result in soluble complexes with iron. The reactivity of soluble metal-organic complexes with ROS differs from metal complexes with inorganic ligands such as hydroxide, chloride, sulfate, and nitrate. Chiorcea-Paquim et al. [40] reviewed the literature on phenolic compounds capable of chelating metals and concluded that complexes could either have anti- or pro-oxidant modes of action that scavenge or produce free radicals, respectively, depending on their structure. For example, Jovanovic et al. [41] found that at pH 7, the Fe(III)bis(gallocatechins) have high oxidation potential, which means that the antioxidant properties of gallocatechins are completely suppressed due to complexation with Fe(III). Walpen et al. [42] reported an automated and highly sensitive flow injection analysis system that quantifies the phenolic antioxidant moieties in dissolved organic matter. Using this system for water-soluble organic carbon in aerosol samples would provide invaluable information on their redox properties.

Complexes of transition metals with certain phenolic compounds can result in electron transfer processes that reduce the metal centers and oxidize the organic ligand. For example, under acidic conditions and a system containing catechol/Fe(III), the complete redox couple for the net reaction shown in eq. (5.34) is as follows:

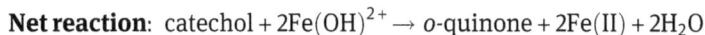

Reduction: $2Fe(OH)^{2+} + 2H^+ + 2e \rightarrow 2Fe(II) + 2H_2O, \quad E_{red} = 0.72\,V \tag{5.32}$

Oxidation: $catechol \rightarrow ortho\text{-quinone} + 2e + 2H^+, \quad E_{ox} = 0.4\,V \tag{5.33}$

Net reaction: $catechol + 2Fe(OH)^{2+} \rightarrow o\text{-quinone} + 2Fe(II) + 2H_2O$

$$E_{net} = E_{red} - E_{ox} = 0.72 - 0.4 = 0.32\,V \tag{5.34}$$

A net positive redox potential, E_{net}, indicates that reaction (5.34) is spontaneous. Similar calculations done for guaiacol and phenol resulted in positive and negative E_{net}, respectively [43]. Two or three polyphenol ligands complexed to iron result in inhibiting the Fe(III) reduction processes [44]. As shown in the next section, spontaneous Fe(III)/Fe(II) cycling with redox-active organic ligands produce ROS in situ, which can lead to secondary formation of soluble and insoluble organic products.

5.1.2.6 Oligomerization and polymerization reactions of organics

Acid-catalyzed reactions take place among the organic and inorganic components in aerosol systems. For example, in the case of carbonyl compounds, Jang et al. [45] reported that these reactions include hydration, hemiacetal/acetal formation, polymerization, and aldol condensation as shown in Scheme 5.2. These reactions were shown to result in a large increase in SOA mass and stabilized organic layers with particle age [45]. In the case of m-xylene-derived SOA, Li et al. [46] showed that increasing the amount of liquid water enhanced the multiphase reactions leading to the formation of oligomers and then to increases in the refractive index, light-scattering efficiency, and direct radiative forcing of the SOA by 20–90%. These results aided in evaluating SOA contribution to regional haze and global radiative balance. Other examples were summarized in the review by Laskin et al. [7].

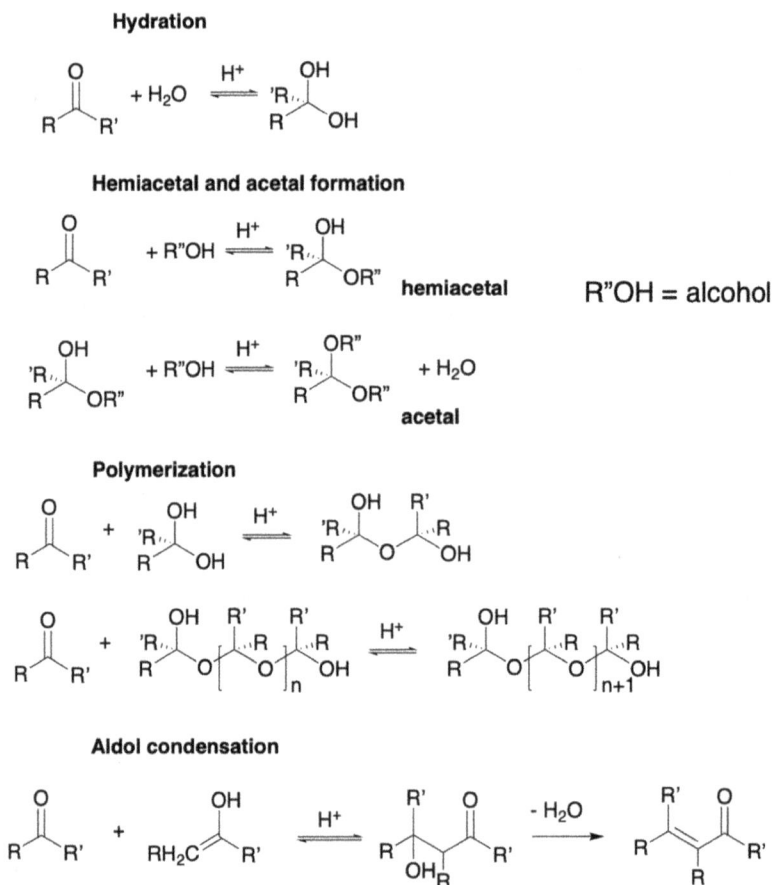

Hydration

Hemiacetal and acetal formation

hemiacetal   R"OH = alcohol

acetal

Polymerization

Aldol condensation

Scheme 5.2: Acid-catalyzed reactions relevant to the carbonyl content in atmospheric aerosol particles as per Jang et al. [45].

Dissolved iron concentrations in rain, fog, snow, and cloud waters were reported to be location dependent and vary by over four orders of magnitude, from 0.1 to 10^3 µM [2], with a typical concentration of dissolved Fe(III) of 1 µM in 20 µm cloud droplets. Using iron chloride as a source of soluble Fe(III) cations to highly aged dust, Al-Abadleh and coworkers showed that soluble iron reacts with aromatic and surface-active phenolic compounds identified in biomass burning organic aerosol extracts, as well as aliphatic dicarboxylic acids commonly formed in volatile organic compound (VOC) oxidation as organic reactants [43]. These reactions efficiently form soluble and insoluble light-absorbing organic products in the dark and under a wide range of atmospheric aerosol mixing states and chemical composition. Also, these reactions were not completely suppressed by the presence of competing iron ligands such as oxalate and sulfate or under UV-irradiated conditions. As reviewed in detail in reference [43], dissolved oxygen is a key ingredient in the reaction mechanism that keeps the Fe(III)/Fe(II) cycle generating in situ ROS such as hydroxyl radicals and hydrogen peroxide.

5.1.3 Effects of ionic strength

The salt content of aerosol liquid water in sea spray aerosols or upon evaporation of water in cloud droplets was found to affect the kinetics of certain reactions [20] and increase or decrease the partitioning/solubility of nonelectrolytes, particularly organics, in processes referred to as "salting in" or "salting out" [47, 48]. The fundamental thermodynamic treatment of salting effects was detailed in references [20, 47]. Equation (5.35) is an empirical equation that describes the impact of inorganic salts on the activity or solubility of neutral organic solutes in aqueous solutions [49]:

$$\log(\gamma/\gamma_0) = \log(S_0/S) = K_S[\text{salt}] = \log\left(K_{1/\text{salt water}}/K_{1/\text{water}}\right) \tag{5.35}$$

where S_0 and S are the solubilities of a solute in pure water and salt solutions, K_S (M^{-1}) is the empirical Setschenow or salting-out coefficient, [salt] (mol L^{-1}) is the salt solution concentration, $K_{1/\text{salt water}}$ and $K_{1/\text{water}}$ are equilibrium partitioning coefficients of an organic solute between the aqueous phase and another nonaqueous phase.

For organics in $(NH_4)_2SO_4$ and sodium chloride solutions, Wang et al. [49] reported measurements and modeling of a series of neutral organic compounds and found that the measured K_S values were all positive in the range 0.216–0.729 M^{-1} and that for the same organic compound, K_S values for $(NH_4)_2SO_4$ are higher than NaCl. The latter result shows that $(NH_4)_2SO_4$ has a higher salting out effect than NaCl. As emphasized by Wang et al. [49], $(NH_4)_2SO_4$ does not only affect the solvation of organic compounds in water, but could influence the reactive fate of some solvated organics as in the case of glyoxal hydration reactions. Other examples that examined the effect of ionic strength in bulk aqueous phase reactions include the oxidation of methoxyphenols by ozone [50], oxidation of benzene by hydroxyl radicals [51], and iron-catalyzed oxidative polymerization of catechol [32].

5.2 Photochemical reactions

The photochemistry of tropospheric aqueous phase has been the subject of extensive reviews [3, 4, 20, 52–57]. Reactive atmospheric chromophores are NO_2^-/NO_3^-, humic-like substances (HULIS), and dissolved iron species that absorb in the near UV light ($\lambda > 300$ nm). The following sections highlight selected examples from recent studies.

5.2.1 Photoformation of ROS from atmospheric chromophores

5.2.1.1 Hydroxyl radicals

Gligorovski et al. [3] reviewed the literature on the reaction pathways that lead to ˙OH radical formation in the aqueous phase, which are listed in Table 5.2. The quantum yields of these reactions are wavelength dependent. The rates of these reactions are also affected by pH. Hydroxyl radicals react with dissolved organic matter in different ways that include double bond addition, hydrogen abstraction, electron transfer, or hydroxylation of aromatic rings [4]. The formation of alkyl (R˙) and alkoxy radicals (RO˙) can result in functionalization/fragmentation reactions leading to decreases in molecular weight and eventual mineralization of the organics. These organic radicals can also lead to dimerization/oligomerization reactions that increase the molecular weight of dissolved organics. The solvent cage effect plays a role in increasing selectivities compared to the aqueous phase [58]. Hydrogen peroxide (H_2O_2) exists in cloud water due to partitioning from the gas phase or the disproportionation reaction of HO_2^- according to the following equation [4, 10]:

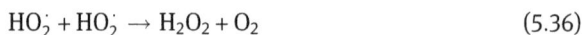

$$HO_2^- + HO_2^- \rightarrow H_2O_2 + O_2 \tag{5.36}$$

H_2O_2 also forms from photosensitized reactions of chromophoric dissolved organic matter (CDOM). The metal-driven reactions that form H_2O_2 are listed in Section 5.1.2.1. In systems containing iron, the photolysis of iron species dominant under acidic conditions such as $FeOH_2^+$ also forms hydroxyl radicals. If Fe(III) is complexed with inorganic or organic ligands, the photolysis of these complexes yields reactive organic radicals that can carry on further reactions [30].

The photochemistry of CDOM is important to understand the photoreactivity of HULIS, especially as a photosensitizer, not only in aqueous solution but also in ice and snow [59, 60]. Excited triplet states, $^3CDOM^\star$, form particularly from the aromatic carbonyls and quinones, upon light absorption of CDOM. These reactive moieties have oxidizing character and can react with oxidizable solutes (S-H) via electron or hydrogen atom abstraction. Chen et al. [61] examined the formation characteristics of ^{3}CDOM* in primary, secondary, and ambient aerosols. Their results showed that organics from biomass combustion have the strongest capacity of ^{3}CDOM* generation, whereas vehicle emissions are the weakest. Ambient aerosols collected during the winter season have higher capacity to generate ^{3}CDOM* than during summer. They also found that the triplet state induces the formation of about third of the ˙OH radicals in these systems. Through structure–activity relationship analysis, they identified C1 and C3 chromophores, which may be N-containing substances, instead of HULIS and phenol-like chromophores, to significantly contribute to ^{3}CDOM* generation. As per the review by Win et al. [57] and Al-Abadleh [30], the chelating abilities of HULIS with transition metals, particularly iron, change their

Table 5.2: List of photochemical reactions that lead to ˙OH radical formation in the aqueous phase as per Gligorovski et al. [3].

Chromophore	Photochemical reactions
NO_3^-	$NO_3^- + hv \rightleftharpoons [O^{\cdot -} + \ ^{\cdot}NO_2]_{cage} \rightarrow O^{\cdot -} + \ ^{\cdot}NO_2$ $O^{\cdot -} + H^+ \rightleftharpoons \ ^{\cdot}OH$ $NO_3^- + hv \rightarrow ONOO^-$ $ONOO^- + H^+ \rightleftharpoons ONOOH$ $ONOOH \rightarrow \ ^{\cdot}OH + \ ^{\cdot}NO_2$
NO_2^-	$NO_2^- + hv \rightleftharpoons O^{\cdot -} + \ ^{\cdot}NO$ $O^{\cdot -} + H^+ \rightleftharpoons \ ^{\cdot}OH$
H_2O_2	$H_2O_2 + hv \rightarrow 2 \ ^{\cdot}OH$ $H_2O_2 + \ ^{\cdot}OH \rightarrow HO_2^{\cdot} + H_2O$
Fe(III)	$FeOH^{2+} + hv \rightarrow Fe(II) + \ ^{\cdot}OH$ $[Fe^{III} - L]^{2+} + hv \rightarrow Fe(II) + L^{\cdot}$ where L is an organic or inorganic ligand with one negative charge.
CDOM	^{1}CDOM $+ hv \rightarrow \ ^1$CDOM$^{\cdot} \rightarrow \ ^3$CDOM$^{\cdot}$ [intersystem crossing (ISC)] ^{1}CDOM $+ hv \rightarrow$ CDOM$^{\cdot +/\cdot -}$ [direct charge transfer (CT)] ^{1}CDOM$^{\cdot} \rightarrow$ CDOM$^{\cdot +/\cdot -}$ [intramolecular CT] ^{3}CDOM$^{\cdot} \rightarrow$ CDOM$^{\cdot +/\cdot -}$ ^{3}CDOM$^{\cdot} + [S - H] \rightarrow$ CDOM$^{\cdot -} + S - H^{\cdot +}$ ^{3}CDOM$^{\cdot} + [S - H] \rightarrow [$CDOM $- H^{\cdot}] + S^{\cdot}$ ^{3}CDOM$^{\cdot} + H_2O/OH^- \rightarrow [$CDOM $- H^{\cdot}]/$CDOM$^{\cdot -} + \ ^{\cdot}OH$ ^{3}CDOM$^{\cdot} + O_2 \rightarrow$ CDOM $+ \ ^1O_2$ (singlet oxygen formation) ^{3}CDOM$^{\cdot} + O_2 \rightarrow$ CDOM $+ O_2$ (excitation energy loss) ^{3}CDOM$^{\cdot} + O_2 \rightarrow$ Products (reactive quenching) ^{3}CDOM$^{\cdot} + O_2 \rightarrow \ \cdot$CDOM$^{\cdot +} + O_2^{\cdot -}$ $[$CDOM $- H^{\cdot}] + O_2 \rightarrow$ CDOM $+ HO_2^{\cdot}$ $HO_2^{\cdot} \rightleftharpoons O_2^{\cdot -} + H^+$ $HO_2^{\cdot} + O_2^{\cdot -} + H^+ \rightarrow H_2O_2 + O_2$ where S $-$ H represents oxidizable solutes

electronic properties and photochemical reactivity. Hence, for multicomponent systems, quantifying the metal content of HULIS would aid in understanding their direct versus photosensitized reactivity.

Given the high reactivity of the $^.$OH radicals and multiple paths that lead to their formation, their role in changing the chemical composition of multicomponent aerosols will depend on the dominant reactions that act as sinks. For example, Anastasio and Newberg [62] reported the sources and sinks of $^.$OH radicals in aqueous extracts of sea salt particles collected from the coast of northern California. Note that the term "sea spray" replaced sea salt particles to accurately reflect the chemical composition of these materials containing inorganic and organic components [63]. Illuminating these extracts formed $^.$OH radicals with rates comparable to those calculated for the partitioning of gas phase $^.$OH radicals to the particles, and 3–4 orders of magnitude greater than the $^.$OH photoformation rates in surface seawater. Lifetimes in the order of 10^{-9}–10^{-8} s were calculated for $^.$OH in these systems. The dominant source of $^.$OH radicals in the particle extracts were found to be from nitrate photolysis, whereas reactions with organics, bromide, and chloride were the dominant sinks of $^.$OH radicals that accounted for 25% of their loss. The authors discussed how particle reactions would probably alter the budgets of gases such as ozone and VOC in the marine boundary layer.

5.2.1.2 Singlet oxygen ($^1O_2$)

Table 5.2 shows the photosensitized formation of $^1O_2$ from the reaction of ^{3}CDOM* with dissolved oxygen. Chen et al. [61] reported that the triplet state induced the formation of at least third of the $^1O_2$ from primary, secondary, and ambient aerosols in China. Manfrin et al. [64] assessed the importance of laboratory-generated SOA from toluene, biphenyl, naphthalene, and 1,8-dimethylnaphthalene and field-collected PM10. Figure 5.3 shows a comparison of the quantum yields for $^1O_2$, $^.$OH, and H_2O_2 as per SOA type, which ranged from 1.2 to 4.5×10^{-2}, 4.6 to 24×10^{-5}, and 1.5 to 5.5×10^{-2}, respectively. The authors emphasized that while $^1O_2$ is a selective oxidant of organic molecules with reaction rates 2 to 3 orders of magnitude smaller than those of the OH radical, they predict that it would be a competitive oxidant to $^.$OH radicals in atmospherically relevant aqueous solutions. This is because the measured $^1O_2$ steady-state concentrations reported in their study were about 3 orders of magnitude larger than those of the $^.$OH radical.

5.2.2 Aqueous pyruvic acid

The bulk photochemistry of pyruvic acid ($pK_a = 2.18$) received much attention because it is the simplest α-keto acid found in the gas and aqueous phases in the atmosphere [65]. This acid undergoes hydration forming its diol form, 2,2-dihydroxypropanoic acid, according to the following equation:

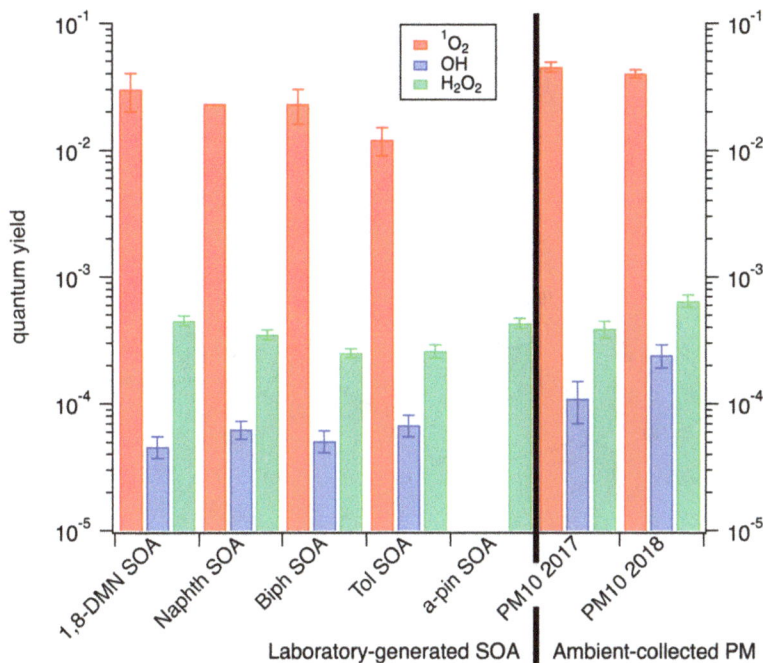

Figure 5.3: Quantum yields of each oxidant measured at 311 nm for laboratory-generated SOA and ambient-collected PM10. Figure and caption were reproduced with permission from reference [64], © The American Chemical Society, 2019.

$$CH_3C(O)CO_2H + H_2O \rightleftharpoons CH_3C(OH)_2CO_2H \qquad (5.37)$$

The equilibrium constant for this reaction was found to be temperature and pH dependent, with a value of 1.54 at pH 2.06 for a 0.1 M concentration [65]. Harris et al. [65] quantified the kinetics of the main reactions involved in the photolysis of pyruvic acid. Figure 5.4 shows a schematic for the main reactions that take place upon light absorption by pyruvic acid in the aqueous phase. Degradation of pyruvic acid by photolysis was reported as a significant pathway compared with ˙OH oxidation and gas-phase photolysis. This degradation pathway results in the formation of higher molecular weight compounds such as dimethyltartaric acid (dimer of pyruvic acid). The dissolved oxygen to pyruvic acid ratio was found to affect the mechanism and hence the photoproduct distribution. Also, the initial pyruvic acid concentration was found to affect the primary reaction pathways. As emphasized by the authors, understanding the fate of pyruvic acid provides insight into the chemistry of other important α-keto acids [65].

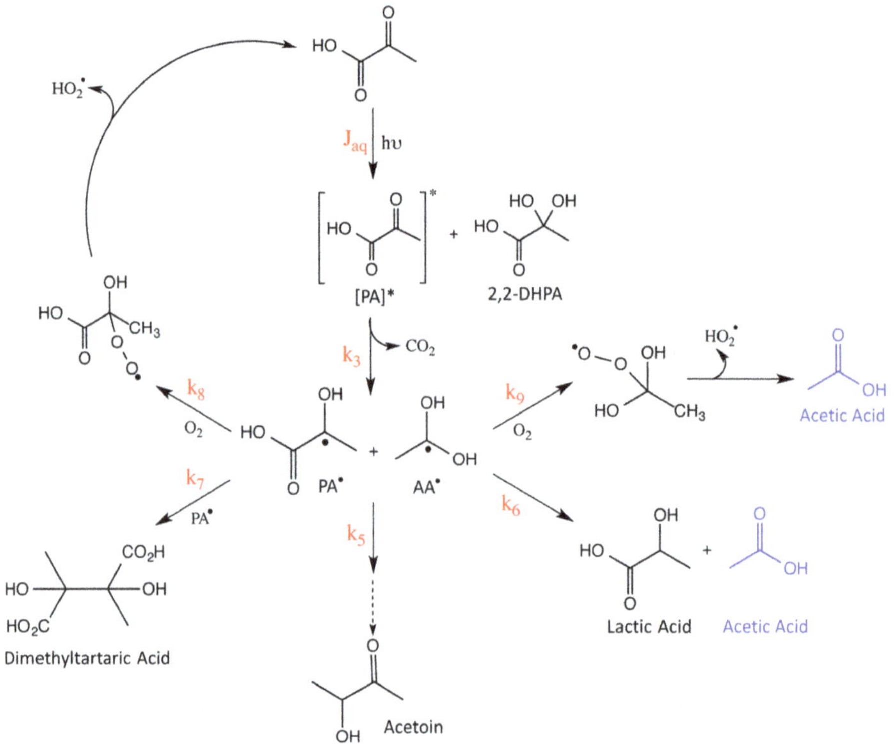

Figure 5.4: Summary of the main reactions involved in the kinetics of the photolysis of pyruvic acid. Rate constants used in the kinetic analysis are shown in red. The label 2,2 DHPA represents 2,2-dihydroxypropanoic acid, the hydrated form of pyruvic acid. Figure and caption were reproduced with permission from reference [65], © The American Chemical Society, 2014.

5.2.3 Aqueous SOA and brown carbon

To simulate photo-oxidation of BrC in atmospheric aqueous droplets, Hems et al. [66] generated BrC in the lab from wood burning and exposed the solutions to UV-B light for comparison with ˙OH oxidation. Reaction products were analyzed using UV–vis spectroscopy, NMR spectroscopy, aerosol chemical ionization mass spectrometry, and liquid chromatography-mass spectrometry (electrospray ionization). Their results showed an increase in the absorbance at 400 nm during ˙OH oxidation followed by a loss in absorbance, which was attributed to initial formation of aromatic dimer compounds and then decomposition to smaller molecules. Exposure to UV-B light resulted in an increase in the absorbance at 400 nm that lasted up to 6 h. The kinetics of the photoinduced loss in absorbance appeared to be slower than the ˙OH oxidation. These results implied that BrC in aqueous aerosol systems is relatively long-lived when considering primarily photo-oxidation aging processes.

Moreover, phenolic carbonyls such as vanillin, acetovanillone, guaiacyl acetone, syringaldehyde, acetosyringone, and coniferyl aldehyde are emitted from wood combustion, and their aqueous photochemistry was investigated by Smith et al. [67]. The authors found that the structure affected the rates of photodegradation and hence atmospheric lifetime that ranged from 13 to 140 min. The direct photodegradation pathways of most of the above phenolic carbonyls were found to be dominant in contrast to aliphatic carbonyls. Guaiacyl acetone was an exception, where direct photodegradation was very slow. This result was explained by its structure where the carbonyl group is not directly connected to the aromatic ring. Triplet excited states of these compounds were reported to be efficient oxidizers of noncarbonyl phenols. Both photodegradation pathways (direct and via triplet excited states) resulted in the formation of low volatility products, with SOA mass yields ranging from 80% to 140%. Smith et al. [67] estimated a 3–5% enhancement in the primary organic aerosol mass from wood combustion due to SOA formation from the direct photodegradation of phenolic carbonyls within a few hours.

The effect of pH on photochemical processing of SOA in the aqueous phase was studied by Amorim et al. [68]. α-Pinene was used to generate SOA, and the water-soluble fraction was used for reactions with $^{\cdot}$OH generated in situ in a photoreactor containing H_2O_2. Figure 5.5 shows the dependency of the second-order rate coefficients with $^{\cdot}$OH, k^{II}_{OH}, and with organic acids extracted from α-pinene SOA: norpinic acid or terpenylic acid ($C_8H_{12}O_4$), *cis*-pinic acid ($C_9H_{14}O_4$), *cis*-limononic acid ($C_{10}H_{16}O_3$), pinyl-diaterpenylic ester ($C_{17}H_{26}O_8$), and isomer *cis*-pinonic acid ($C_{10}H_{16}O_4$). $C_{10}H_{16}O_5$ was assigned to a potential ROOH compound. While the organic acids exhibited larger OH reactivities at pH 10, values of k^{II}_{OH} show less pH dependency compared with formic acid, which is used as a reference compound, with $k^{II}_{OH} \sim 8 \times 10^8$ M^{-1} s^{-1} (pH 2) versus 2×10^{10} M^{-1} s^{-1} (pH 10). These differences were attributed to the reduction in the relative importance of electron transfer reactions in organic acids with larger carbon chain (#C atoms > 8). These results indicate that simplification of cloud chemistry models could be done when accounting for structure–reactivity relationships.

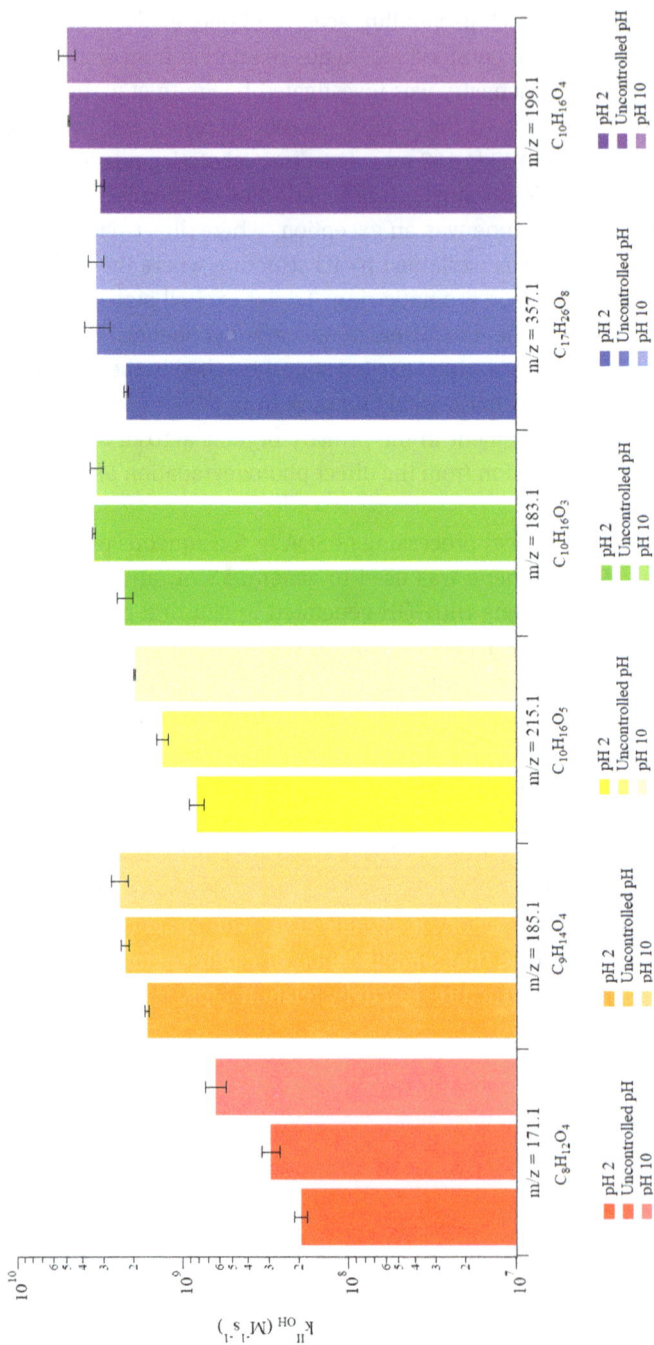

Figure 5.5: OH reactivities of α-pinene SOA OAs at uncontrolled pH, pH 2, and pH 10. Figure and caption were reproduced with permission from reference [68], © The American Chemical Society, 2019.

5.2.4 Synergetic reactions: sulfate formation and nitrate reduction

Gen et al. [29] explored the photochemistry of aqueous iron chloride species found in aerosol liquid water of particulate iron chloride. Figure 5.6 shows the series of reactions initiated by photochemical formation of $^{\cdot}Cl$ and $Cl_2^{\cdot-}$ from $FeCl^{2+}$ and $FeCl_2^{+}$ species and they are linked to gas phase SO_2 uptake and subsequent oxidation to sulfate. Through a series of control experiments that examined the role of light and dissolved oxygen, dark reactions and those conducted using nitrogen gas had lower rates of sulfate production by 10×. Replacing chloride with nitrate also lowered the rate of sulfate production with irradiation using the same light source (xenon lamp) by nearly 10×. The addition of organic ligands such as oxalate and malonate was also investigated and found to result in an initial delay in sulfate production that increased in length with increasing initial concentration of the organics. An increase in sulfate production was observed after that delay to the same levels as in the absence of the organics. Overall, the experiments and modeling conducted in this study found that while there is a poor correlation between the reactive uptake coefficient of SO_2 (γSO_2) and iron chloride concentration, a good positive correlation between γSO_2 and the $^{\cdot}Cl$ production rate was obtained highlighting the latter in sulfate formation. Competing reactions such as

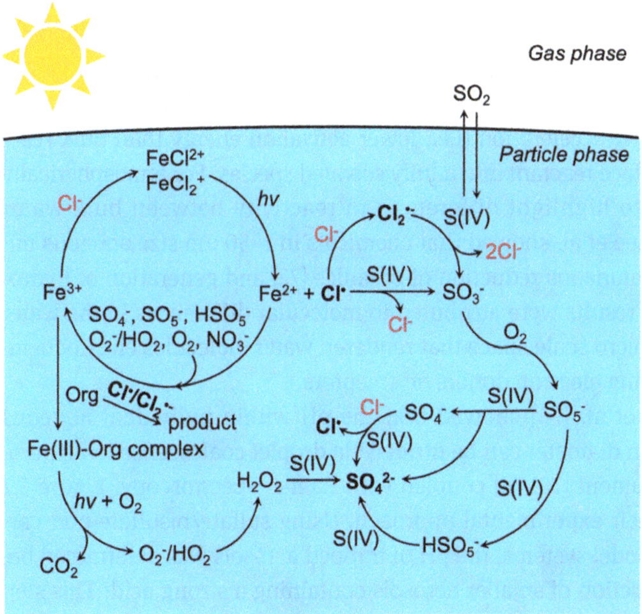

Figure 5.6: Simplified schematic of the proposed reaction mechanisms for the role of iron chloride photochemistry in SO_2 gas-phase uptake and sulfate production. Figure and caption were reproduced with permission from reference [68], © The American Chemical Society, 2020.

chlorination of organics might lower ˙Cl levels, which warrants further investigations relative to other pathways that lead to sulfate production.

Since aqueous Fe(II) is dominant in the aqueous phase during daytime, Gen et al. [33] also reported the formation of N(III) (HONO/NO$_2^-$) from photochemical reactions in micrometer-sized droplets containing iron–organic complexes and nitrate. N(III) species were detected indirectly through their reaction with dissolved SO$_2$ leading to sulfate formation. They observed an increase in sulfate production rate with the concentration of iron–organic complexes in nitrate particles. Higher concentrations of iron–organic complexes were found to yield higher nitrate decay rates. This study highlighted the synergistic effect between iron–organic complexes and nitrate under irradiation as a source of N(III), particularly HONO, which upon photolysis influences hydroxyl radicals, the most important daytime oxidant in the atmosphere.

5.3 Chemistry in microdroplets

Selected reactions in aqueous microdroplets were reported to proceed at much faster rates than in bulk water [69, 70]. Quantum mechanical modeling of the reaction between phenylhydrazine and indoline-2,3-dione with explicit solvent calculations was chosen to reproduce experimental data that showed a factor of 10^4 increase in reaction rate in microdroplets using HCl/methanol [71]. The calculations showed that molecular orientation of protonated phenylhydrazine, which placed the charge sites at the surface of the droplet, had high energy for being partly solvated. The reaction pathway that resulted in strong acceleration (i.e., lower activation energy than bulk reaction) transformed the surface reactant into a fully solvated species. For atmospherically relevant reactions and to highlight differences in reactivity between bulk water and micro-sized water, Lee et al. showed that chemicals in 1–50 μm size aqueous microdroplets induced spontaneous reduction of organics [72] and generation of hydrogen peroxide [73]. These results were attributed to molecular differences in the water structure in a confined micro-scale space that rendered water molecules electrochemically active without adding electron donors or acceptors.

Moreover, Coddens et al. [74] showed that the pH within individual aqueous aerosols that are ~8 μm in diameter can be titrated via droplet coalescence in an aerosol optical tweezer instrument [75, 76] coupled with Raman spectroscopy. Figure 5.7 shows a schematic of their experimental approach. Using sulfate/bisulfate and carbonate/bicarbonate as model systems, the pH of trapped aerosols was determined before and after the introduction of smaller aerosols containing a strong acid. This step opens the door for future studies aimed at understanding pH-dependent reactions and processes relevant to aerosol systems. For example, the formation of organosulfate esters from isoprene was reported by Bondy et al. [77] in Raman microspectroscopy experiments using micron-sized droplets deposited on quartz substrates from nebulizing aqueous solutions with known concentrations of the organic standards.

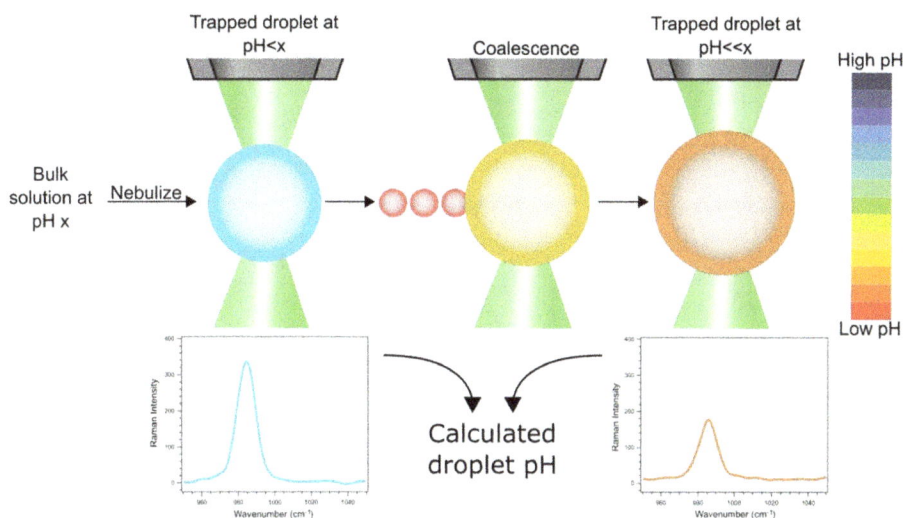

Figure 5.7: Schematic of the aerosol-trapping process from bulk solutions of pH x. The initial pH of the aerosol is less than that of the bulk solution. The aerosol pH can be changed through coalescence with another aerosol at a different pH. Raman spectra from the trapped particle are used in conjunction with calibration curves to determine concentration and ultimately calculate the following aerosol pH. Figure and caption were reproduced with permission from reference [74], © The American Chemical Society, 2020.

These lab analyses were complemented with density functional theory calculations and characterizations of ambient particles from a rural forested region with high isoprene emissions to identify isoprene-derived organosulfate Raman signature. The vibrational mode assignments for 2-methylglyceric acid sulfate ester and 3-methyltetrol sulfate esters, along with their hydrolysis products 2-methylglyceric acid and 2-methyltetrols shown in Figure 5.8 were reported. The characterization done in this study aids in future identification of these species within individual SOA particles to assess their abundance and atmospheric implications.

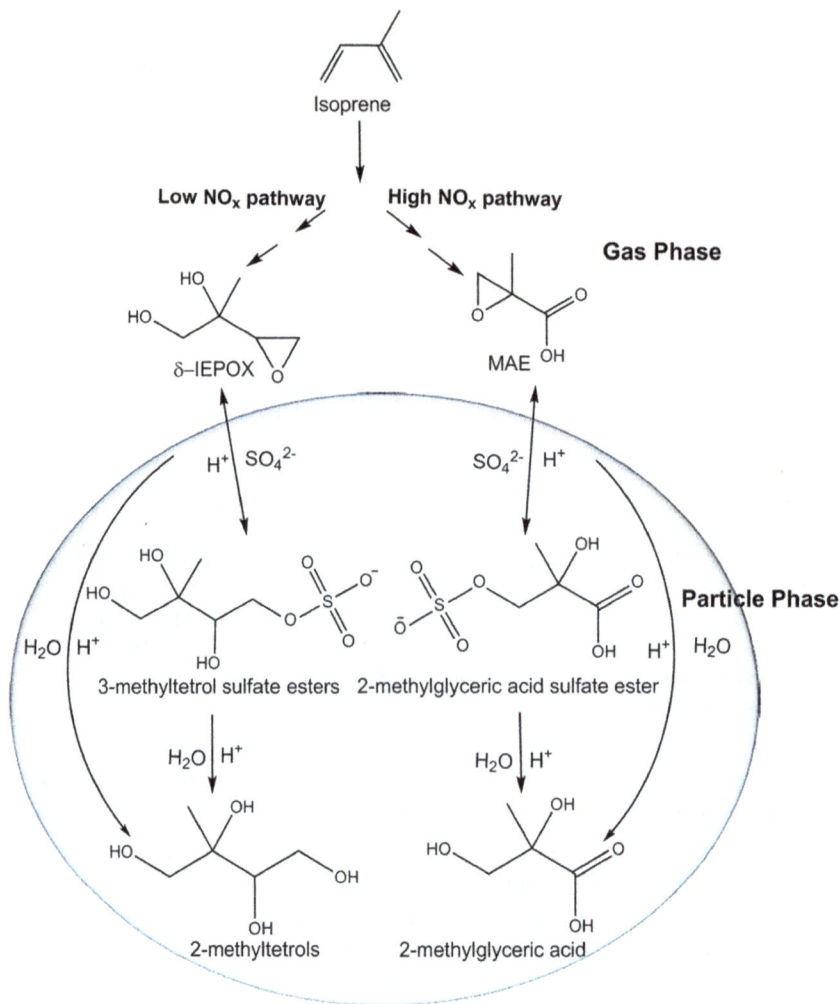

Figure 5.8: Scheme leading to the formation of isoprene-derived SOA compounds: 3-methyltetrol sulfate esters, 2-methyltetrols, 2-methylglyceric acid sulfate ester, and 2-methylglyceric acid. Both the high- and low-NO_x pathways are shown. For simplicity, only one isomer of each respective compound is shown. 2-Methylglyceric acid, 2-methylglyceric acid sulfate ester, 2-methyltetrols, and the 3-methyltetrol sulfate esters are present in the particle phase, while the epoxides are in the gas phase. Figure and caption were reproduced with permission from reference [77], © The American Chemical Society, 2018.

References

[1]   Tilgner A, Schaefer T, Alexander B, Barth MC, Collett JL Jr, Fahey KM, et al. Acidity and the multiphase chemistry of atmospheric aqueous particles and clouds. Atmos Chem Phys. 2021;21:13483–13536.

[2]   Deguillaume L, Leriche M, Desboeufs K, Mailhot G, George C, Chaumerliac N. Transition metals in atmospheric liquid phases: sources, reactivity, and sensitive parameters. Chem Rev. 2005;105(9):3388–3431.

[3]   Gligorovski S, Strekowski R, Barbati S, Vione D. Environmental implications of hydroxyl radicals ($\cdot$OH). Chem Rev. 2015;115:13051–13092.

[4]   Bianco A, Passananti M, Brigante M, Mailhot G. Photochemistry of the cloud aqueous phase: A review. Molecules. 2020;25:423, doi:10.3390/molecules25020423.

[5]   Ervens B. Progress and Problems in Modeling Chemical Processing in Cloud Droplets and Wet Aerosol Particles. In: Hunt SW, Laskin A, Nizkorodov SA, editors. Multiphase Environmental Chemistry in the Atmosphere. Washington DC: American Chemical Society; 2018. pp. 327–345.

[6]   de Semainville PG, D'Anna B, George C. Aqueous phase reactivity of nitrate radicals (NO_3) toward dicarboxylic acids. Z Phys Chem. 2010;224:1247–1260.

[7]   Laskin A, Laskin J, Nizkorodov SA. Chemistry of atmospheric brown carbon. Chem Rev. 2015;115(10):4335–4382.

[8]   Lin P, Laskin J, Nizkorodov SA, Laskin A. Revealing brown carbon chromophores produced in reactions of methylglyoxal with ammonium sulfate. Environ Sci Technol. 2015;49(24):14257–14266.

[9]   Lee AKY, Zhao R, Li R, Liggio J, Li S-M, Abbatt JPD. Formation of light absorbing organo-nitrogen species from evaporation of droplets containing glyoxal and ammonium sulfate. Environ Sci Technol. 2013;47(22):12819–12826.

[10]   Buxton GV, Salmon GA. On the chemistry of inorganic free radicals in cloud water. Prog React Kinet Mech. 2003;28:257–297.

[11]   Siefert R, Johansen AM, Hoffmann MR, Pehkonen SO. Measurements of trace metal (Fe, Cu, Mn, Cr) oxidation states in fog and stratus clouds. J Air Waste Manage Assoc. 1998;48(2):128–143.

[12]   Al Nimer A, Rocha L, Rahman MA, Nizkorodov SA, Al-Abadleh HA. Effect of oxalate and sulfate on iron-catalyzed secondary brown carbon formation. Environ Sci Technol. 2019;53(12):6708–6717.

[13]   Kremer ML. The Fenton Reaction. Dependence of the Rate on pH. J Phys Chem A. 2003;107:1734–1741.

[14]   Duesterberg CK, Waite TD. Kinetic modeling of the oxidation of p-hydroxybenzoic acid by Fenton's reagent: Implications of the role of quinones in the redox cycling of iron. Environ Sci Technol. 2007;41(11):4103–4110.

[15]   Chang C-Y, Hsieh Y-H, Cheng K-Y, Hsieh -L-L, Cheng T-C, Yao K-S. Effect of pH on Fenton process using estimation of hydroxyl radical with salicylic acid as trapping reagent. Water Sci Technol. 2008;58(4):873–879.

[16]   Meyerstein D. Re-examining Fenton and Fenton-like reactions. Nature Rev Chem. 2021;5:595–597.

[17]   Paulson SE, Gallimore PJ, Kuang XM, Chen JR, Kalberer M, Gonzalez DH. A light-driven burst of hydroxyl radicals dominates oxidation chemistry in newly activated cloud droplets. Sci Adv. 2019;5(1–7):eaav7689.

[18] Fang T, Lakey PSJ, Rivera-Rios JC, Keutsch FN, Shiraiwa M. Aqueous-phase decomposition of isoprene hydroxy hydroperoxide and hydroxyl radical formation by Fenton-like reactions with iron ions. J Phys Chem A. 2020;124(25):5230–5236.

[19] Rabani J, Klug-Roth D, Lilie J. Pulse radiolytic investigations of the catalyzed disproportionation of peroxy radicals. Aqueous Cupric Ions J Phys Chem. 1973;77(9): 1169–1175.

[20] Herrmann H, Schaefer T, Tilgner A, Styler SA, Weller C, Teich M, et al. Tropospheric aqueous-phase chemistry: kinetics, mechanisms, and its coupling to a changing gas phase. Chem Rev. 2015;115(10):4259–4334.

[21] Harris E, Sinha B, van Pinxteren D, Tilgner A, Fomba KW, Schneider J, et al. Enhanced role of transition metal ion catalysis during in-cloud oxidation of SO_2. Science. 2013;340:727–730.

[22] Fronaeus S, Berglund J, Elding LI. Iron-manganese redox processes and synergism in the mechanism for manganese-catalyzed autoxidation of hydrogen sulfite. Inorg Chem. 1998;37: 4939–4944.

[23] Bruggemann M, Xu R, Tilgner A, Kwon KC, Mutzel A, Poon HY, et al. Organosulfates in ambient aerosol: State of knowledge and future research directions on formation, abundance, fate, and importance. Environ Sci Technol. 2020;54:3767–3782.

[24] Huang L, Coddens EM, Grassian VH. Formation of organosulfur compounds from aqueous phase reactions of S(IV) with methacrolein and methyl vinyl ketone in the presence of transition metal ions. ACS Earth Space Chem. 2019;3:1749–1755.

[25] Huang L, Liu T, Grassian VH. Radical-initiated formation of aromatic organosulfates and sulfonates in the aqueous phase. Environ Sci Technol. 2020;54:11857–11864.

[26] Buresh RJ, Moraghan JT. Chemical Reduction of Nitrate by Ferrous Iron. J Environ Qual. 1976;5:320–325.

[27] Ottley CJ, Davison W, Edmunds WM. Chemical catalysis of nitrate reduction by iron(II). Geochem Cosmochim Acta. 1997;61:1819–1828.

[28] Cwiertny DM, Baltrusaitis J, Hunter GJ, Laskin A, Scherer MM, Grassian VH. Characterization and acid-mobilization study of iron-containing mineral dust source materials. J Geophys Res-Atmos. 2008;113:D05202, doi:10.1029/2007jd009332.

[29] Gen M, Zhang R, Li Y, Chan CK. Multiphase photochemistry of iron-chloride containing particles as a source of aqueous chlorine radicals and its effect on sulfate production. Environ Sci Technol. 2020;54(16):9862–9871.

[30] Al-Abadleh HA. A review on the bulk and surface chemistry of iron in atmospherically-relevant systems containing humic like substances. RSC Adv. 2015;5(57):45785–45811.

[31] Visual MINTEQ ver. 3.1. [Available from: https://vminteq.lwr.kth.se.

[32] Al-Abadleh HA, Rana MS, Mohammed W, Guzman MI. Dark iron-catalyzed reactions in acidic and viscous aerosol systems efficiently form secondary brown carbon. Environ Sci Technol. 2021;55(1):209–219.

[33] Gen M, Zhang R, Chan CK. Nitrite/nitrous acid generation from the reaction of nitrate and Fe (II) Promoted by photolysis of iron–organic complexes. Environ Sci Technol. 2021, doi:doi. org/10.1021/acs.est.1c05641.

[34] Mentasti E, Baiocchi C. The equilibria and kinetics of the complex formation between Fe(III) and tartaric and citric acids. J Coord Chem. 1980;10(4):229–237.

[35] Moorhead EG, Sutin N. Rate and equilibrium constants for the formation of the monooxalate complex of iron(III). Inorg Chem. 1966;5(11):1866–1871.

[36] Kormanyos B, Peintler G, Nagy A, Nagypal I. Peculiar kinetics of the complex formation in the iron(III)–sulfate system. Int J Chem Kin. 2008;40(3):114–124.

[37] Mentasti E. Equilibria and kinetics of the complex formation between iron(III) and α-hydroxycarboxylic acids. Inorg Chem. 1979;18(6):1512–1515.

[38] Mentasti E, Pelizzetti E. Reactions between iron(III) and catechol (o-dihydroxybenzene). part I. Equilibria and kinetics of complex formation in aqueous acid solution. J Chem Soc, Dalton Trans. 1973;1(23):2605–2608.

[39] Tran A, William G, Younus S, Ali NN, Blair SL, Nizkorodov SA, et al. Efficient formation of light-absorbing polymeric nanoparticles from the reaction of soluble Fe(III) with C4 and C6 dicarboxylic acids. Environ Sci Technol. 2017;51(17):9700–9708.

[40] Chiorcea-Paquim AM, Enache TA, Gil ED, Oliveira-Brett AM. Natural phenolic antioxidants electrochemistry: Towards a new food science methodology. Comrr Rev Food Sci F. 2020;19: 1680–1726.

[41] Jovanovic SV, Simic MG, Steenken S, Hara Y. Iron complexes of gallocatechins. Antioxidant action or iron regulation? J Chem Soc Perkin Trans. 1998;2:2365–2369.

[42] Walpen N, Schroth MH, Sander M. Quantification of phenolic antioxidant moieties in dissolved organic matter by flow-injection analysis with electrochemical detection. Environ Sci Technol. 2016;50:6423–6432.

[43] Al-Abadleh HA. Aging of atmospheric aerosols and the role of iron in catalyzing brown carbon formation. Environ Sci: Atmos. 2021;1:297–345.

[44] Perron NR, Brumaghim JL. A review of the antioxidant mechanisms of polyphenol compounds related to iron binding. Cell Biochem Biophys. 2009;53(1):75–100.

[45] Jang M, Czoschke NM, Lee S, Kamens RM. Heterogeneous atmospheric aerosol production by acid-catalyzed particle-phase reactions. Science. 2002;298(5594):814–817.

[46] Li K, Li J, Liggio J, Wang W, Fe M, Liu Q, et al. Enhanced light-scattering of secondary organic aerosols by multiphase reactions. Environ Sci Technol. 2017;51(3):1285–1292.

[47] Grover PK, Ryall RL. Critical appraisal of salting-out and its implications for chemical and biological sciences. Chem Rev. 2005;105(1):1–10.

[48] You Y, Smith ML, Song M, Martin ST, Bertram AK. Liquid–liquid phase separation in atmospherically relevant particles consisting of organic species and inorganic salts. Int Rev Phys Chem. 2014;33(1):43–77.

[49] Wang C, Lei YD, Endo S, Wania F. Measuring and modeling the salting-out effect in ammonium sulfate solutions. Environ Sci Technol. 2014;48(22):13238–13245.

[50] Mekic M, Wang Y, Loisel G, Vione D, Gligorovski S. Ionic strength effect alters the heterogeneous ozone oxidation of methoxyphenols in going from cloud droplets to aerosol deliquescent particles. Environ Sci Technol. 2020;54(20):12898–12907.

[51] Heath AA, Valsaraj KT. Effects of temperature, oxygen level, ionic strength, and pH on the reaction of benzene with hydroxyl radicals at the air–water interface in comparison to the bulk aqueous phase. J Phys Chem A. 2015;119(31):8527–8536.

[52] Vione D, Maurino V, Minero C, Pelizzetti E, Harrison MAJ, Olariu R-I, et al. Photochemical reactions in the tropospheric aqueous phase and on particulate matter. Chem Soc Rev. 2006;35:441–453.

[53] Vione D. Photochemical Transformation Processes of Environmental Significance. In: Pignataro B, editor. Tomorrow's Chemistry Today. Weinheim: Wiley-VCH; 2008. pp. 429–453.

[54] Lueder U, Jorgensen BB, Kappler A, Schmidt C. Photochemistry of iron in aquatic environments. Environ Sci: Processes Impacts. 2020;22(12–24).

[55] George C, D'Anna B, Herrmann H, Weller C, Vaida V, Donaldson DJ, et al. Emerging Areas in Atmospheric Photochemistry. In: McNeill VF, Ariya PA, editors. Atmospheric and Aerosol Chemistry, vol. 339, Heidelberg: Springer; 2012. pp. 1–54.

[56] Hoffmann MR. Homogeneous and Heterogeneous Photochemistry in the Troposphere. In: Boule P, Bahnemann DW, Robertson P, editors. Environmental Photochemistry Part II. Handb. Environ. Chem. Ser. 2M. Berlin: Springer; 2005. pp. 77–118.

[57] Win MS, Tian Z, Zhao H, Xiao K, Peng J, Shang Y, et al. Atmospheric HULIS and its ability to mediate the reactive oxygen species (ROS): A review. J Environ Sci. 2018;71:13–31.

[58] Kopinke F-D GA. What controls selectivity of hydroxyl radicals in aqueous solution? indications for a cage effect. J Phys Chem A. 2017;121:7947–7955.

[59] Grannas AM, Jones AF, Dibb J, Ammann M, Anastasio C, Beine HJ, et al. An overview of snow photochemistry: Evidence, mechanisms and impacts. Atmos Chem Phys. 2007;7:4329–4373.

[60] Grannas AM, Pagano LP, Pierce BC, Bobby R, Fede A. Role of dissolved organic matter in ice photochemistry. Environ Sci Technol. 2014;48:10725–10733.

[61] Chen Q, Mu Z, Xu L, Wang M, Wang J, Shan M, et al. Triplet-state organic matter in atmospheric aerosols: Formation characteristics and potential effects on aerosol aging. Atmos Environ. 2021;252(1–11):118343.

[62] Anastasio C, Newberg JT. Sources and sinks of hydroxyl radical in sea-salt particles. J Geophys Res. 2007;112:D10306, doi:10.1029/2006JD008061.

[63] Bertram TH, Cochran RE, Grassian VH, Stone EA. Sea spray aerosol chemical composition: Elemental and molecular mimics for laboratory studies of heterogeneous and multiphase reactions. Chem Soc Rev. 2018;47:2374–2400.

[64] Manfrin A, Nizkorodov SA, Malecha KT, Getzinger GJ, McNeill K, Borduas-Dedekind N. Reactive oxygen species production from secondary organic aerosols: The importance of singlet oxygen. Environ Sci Technol. 2019;53:8553–8562.

[65] Harris AER, Ervens B, Shoemaker RK, Kroll JA, Rapf RJ, Griffith EC, et al. Photochemical kinetics of pyruvic acid in aqueous solution. J Phys Chem A. 2014;118:8505–8516.

[66] Hems RF, Schnitzler EG, Bastawrous M, Soong R, Simpson AJ, Abbatt JPD. Aqueous photoreactions of wood smoke brown carbon. ACS Earth Space Chem. 2020;4(7):1149–1160.

[67] Smith SD, Kinney H, Anastasio C. Phenolic carbonyls undergo rapid aqueous photodegradation to form low-volatility, light-absorbing products. Atmos Environ. 2016;126: 36–44.

[68] Amorim JV, Wu S, Klimchuk K, Lau C, Williams FJ, Huang Y, et al. pH dependence of the OH reactivity of organic acids in the aqueous phase. Environ Sci Technol. 2020;54(19): 12484–12492.

[69] Arnaud CH. Microdroplets rev up chemical reactions. Chem Eng News. 2017;95(46):16–18.

[70] Banerjee S, Gnanamani E, Yan X, Zare RN. Can all bulk-phase reactions be accelerated in microdroplets? Analyst. 2017;142:1399–1402.

[71] Narendra N, Chen X, Wang J, Charles J, Cooks RG, Kubis T. Quantum mechanical modeling of reaction rate acceleration in microdroplets. J Phys Chem A. 2020;124:4984–4989.

[72] Lee JK, Samanta D, Nam HG, Zare RN. Micrometer-sized water droplets induce spontaneous reduction. J Am Chem Soc. 2019;141:10585–10589.

[73] Lee JK, Walker KL, HAn HS, Kang J, Prinz FB, Waymouth RM, et al. Spontaneous generation of hydrogen peroxide from aqueous microdroplets. PNAS. 2019;116(39):19294–19298.

[74] Coddens EM, Angle KJ, Grassian VH. Titration of aerosol pH through droplet coalescence. J Phys Chem Lett. 2019;10:4476–4483.

[75] Sullivan RC, Boyer-Chelmo H, Gorkowski K, Beydoun H. Aerosol optical tweezers elucidate the chemistry, acidity, phase separations, and morphology of atmospheric microdroplets. Acc Chem Res. 2020;53(11):2498–2509.

[76] Knox KJ. Light-induced Processes in Optically-tweezed Aerosol Droplets. Heidelberg: Springer; 2011.

[77] Bondy AL, Craig RL, Zhang Z, Gold A, Surratt JD, Ault AP. Isoprene-derived organosulfates: Vibrational mode analysis by Raman spectroscopy, acidity-dependent spectral modes, and observation in individual atmospheric particles. J Phys Chem A. 2018;122:303–315.

Index

https://doi.org/10.1515/9781501519376-006

www.ingramcontent.com/pod-product-compliance
Lightning Source LLC
Chambersburg PA
CBHW061417210326
41598CB00035B/6251